Alex
Basics of Engineering and
Technology with Natural Rubber

Rosamma Alex

Basics of Engineering and Technology with Natural Rubber

HANSER

Print-ISBN: 978-1-56990-455-8
E-Book-ISBN: 978-1-56990-950-8

Bibliographic information of the German National Library:
The German National Library lists this publication in the German National Bibliography; detailed bibliographic data are available on the Internet at http://dnb.d-nb.de.

© 2026 Carl Hanser Verlag GmbH & Co. KG, Munich
Vilshofener Straße 10 | 81679 Munich | info@hanser.de
www.hanserpublications.com
www.hanser-fachbuch.de
Editor: Dr. Mark Smith
Production Management: Cornelia Speckmaier
Cover concept: Marc Müller-Bremer, www.rebranding.de, Munich
Cover design: Max Kostopoulos
Cover picture: © stock.adobe.com/美緒奈 谷岡
Typesetting: le-tex publishing services GmbH, Leipzig

The Authors

The Editor

Dr. Rosamma Alex is a leading expert in rubber technology, with a special focus on natural rubber. She served for over 36 years at the Rubber Research Institute of India (RRII) under the Rubber Board, Ministry of Commerce, Government of India, making significant contributions to the advancement of rubber processing and materials science. She retired in 2015 as Joint Director (Rubber Technology) and has since continued to offer her expertise as a consultant in the field.

Dr. Alex has authored more than 40 research articles in scientific journals and presented over 50 papers at national and international conferences. She contributed chapters to two books:

- *Natural Rubber: Agromanagement and Crop Processing*, edited by P. J. George and C. Kuruvilla Jacob (Rubber Research Institute of India), and

- *Rubber Nanocomposites: Preparation, Properties, and Applications*, edited by Dr. Sabu Thomas and Dr. Ranimol Stephen (Wiley).

She has guided five Ph.D. students affiliated with Cochin University of Science and Technology and Mahatma Gandhi University, India.

Dr. Alex's work has earned several recognitions, including the Modi Rubber Prize (1985) from IIT Kharagpur for the best M.Tech thesis, and the IRMRA Best Technical Paper Award in both 2005 and 2012. She holds one U.S. patent and has filed two Indian patents in the area of rubber processing.

She was selected for the prestigious Korea Brain Pool Fellowship, under which she conducted post-doctoral research at Chonbuk National University, South Korea, in 2004. She also completed a short training program at the University of North London,

UK, in 1998. Dr. Alex earned her Ph.D. in Rubber Technology from the Indian Institute of Technology Kharagpur in 1992.

Her personal interests include reading and music.

The Contributors

Pradeepkumar P. Joy

Pradeepkumar P. Joy has over 23 years of expertise in rubber processing and product manufacturing technology. He holds an M.Tech in Polymer Technology from Cochin University of Science and Technology (CUSAT), Kerala, and is a graduate chemical engineer. Since 2007, he has been a faculty member at the National Institute for Rubber Training (NIRT), Rubber Board, India. His prior experience includes working with M/s. Kurlon Limited, Bangalore, a leader in rubberized coir mattresses. He also holds a Postgraduate Diploma in Rubber Technology (PGDIRI) from the Indian Rubber Institute, a Diploma in Polymer Technology from Govt. Polytechnic College, Kottayam, and a Diploma in Statistical Quality Control (SQC) from the Indian Statistical Institute, Karnataka. Currently, he is pursuing a Ph.D. in Chemistry from Mahatma Gandhi University, Keralarrently, he is pursuing a Ph.D. in Chemistry from Mahatma Gandhi University, Kerala.

Sasidharan K. K.

Dr. Sasidharan K. K. has over 37 years of extensive experience in the rubber industry. He began his career as an R&D Chemist at Philips Carbon Black in Kochi, Kerala, where he worked for four years. In 1986, he joined the Government Polytechnic College, Kerala, as an instructor in polymer technology, before moving to the Rubber Research Institute of India (RRII) under the Rubber Board, Government of India. He served at RRII until his retirement in 2019 and has since been active as a consultant in rubber technology.

Dr. Sasidharan has authored 34 journal publications and written two books in the field of rubber technology. He is the recipient of a U.S. patent for his work on latex carbon black masterbatch, and was awarded the IRMRA Best Technical Paper Award in January 2012.

He earned his B.Tech in Polymer Science and Rubber Technology in 1999 and his M. Tech in Polymer Technology in 2002, both from Cochin University of Science and Technology (CUSAT). He received his Ph.D. in Rubber Technology from the same university in 2014.

His interests include reading and photography.

K. N. Madhusoodanan

K. N. Madhusoodanan has over 30 years of experience in scientific research, serving as a scientist at the Rubber Research Institute of India since 1992. His work has made notable contributions to various areas of rubber technology, with a particular focus on latex technology, rubber recycling, and materials science. He has authored more than 15 publications in peer-reviewed journals and industry magazines.

His recent research has explored the reinforcement of natural rubber using polymeric fillers, a subject of significant industrial relevance for improving performance and sustainability in rubber products.

Mr. Madhusoodanan holds an M.Sc. in Chemistry from Mahatma Gandhi University. Beyond his scientific career, he is passionate about photography, and enjoys singing and painting in his leisure time.

Baby Kuriakose

Dr. Baby Kuriakose has over 40 years of experience in the rubber industry. He began his career in 1977 as a scientist at the Rubber Research Institute of India (RRII) under the Rubber Board, Government of India, and rose to the position of Joint Director (Rubber Technology). After retiring from the Rubber Board, he worked as Chief Technical Production Manager at a latex centrifuging unit in Thailand for three years. He has also worked as a consultant in the field of tire retreading.

Dr. Kuriakose has published more than 45 research papers in journals and presented over 50 papers at national and international conferences. He has contributed chapters to two books: Natural Rubber: Agromanagement and Crop Processing and Natural Rubber (published by Rubber Research Institute of India, edited by P. J. George and C. Kuruvilla Jacob) and Biology, Cultivation and Technology (published by Rubber Research Institute of India, edited by N. M. Mathew and M. R. Sethuraj). He has guided six Ph.D. students from Cochin University of Science and Technology and Mahatma Gandhi University, India.

He received the IRMRA Best Technical Paper Award in December 2005 for his work in rubber technology.

Dr. Kuriakose holds an M.Sc. in Chemistry from Kerala University and earned his Ph.D. in Rubber Technology from the Indian Institute of Technology, Kharagpur in 1986.

His interests include reading and photography.

Umasankar G

Umasankar G, a mechanical engineer with over 15 years of experience in the rubber sector, holds a B.Tech degree in Mechanical Engineering from Calicut University and a postgraduate degree in Business Administration from Mahatma Gandhi University,

Kottayam. An expert in rubber processing and engineering, he has been with the Rubber Board of India since 2008. He began his career in the Processing and Product Development Department, contributing to various innovations in primary natural rubber processing. He was later promoted to Factory Manager of the Model TSR Factory.

His experience also extends to the oil and gas sector, where he worked as a Project Planning Engineer in the Middle East. He is currently the Deputy Director (Engineering), Rubber Board. In addition, he serves as a trainer, consultant, and knowledge provider in the rubber processing sector. He is also an active member of two subcommittees under the Bureau of Indian Standards (BIS), contributing to the formulation and revision of Indian standards.

Acknowledgments

I express my sincere gratitude to Dr. Sarwar Dhananiya, former Chairman of the Rubber Board, Government of India, Kottayam, Kerala and Mr. M. Vasanthagesan, Indian Revenue Service Officer, Executive Director of the Rubber Board for graciously writing the foreword for this book.

As the editor, this work would not have been possible without the valuable contributions and collaboration of my co-authors. I deeply appreciate their assistance and subject expertise. I also acknowledge the support and encouragement from fellow scientists and officers at NIRT, the Rubber Board, and my colleagues at the Rubber Research Institute of India (RRII), Rubber Board.

I extend my special thanks to Mr. S. C. Sajayan and Ms. Jayasree T, for their help in providing the necessary support for this book. I am also grateful to Abdulkalam T. M., Environmental Engineer with the Pollution Control Board, Kerala, for his efforts in creating the figures. I appreciate Dr. Manoj Kurian, Scientist, RRII, for supplying valuable technical insights during the writing process.

Furthermore, I sincerely thank Dr. Alex Mathew, Engineer, NVIDIA, Austin, Texas, USA, for his specialized expertise in deriving mathematical equations related to vibration and shock transmissibility, vibration isolation, and energy dissipation in damping materials and rubber.

I am deeply indebted to my mother, husband and family for their unwavering support and encouragement throughout this endeavor. Their love and understanding have been my greatest strength.

Finally, I extend my heartfelt gratitude to my research guide, Prof. S. K. De, for his motivation and guidance, which have inspired me to explore various aspects of rubber technology at the Indian Institute of Technology, Kharagpur, India.

Dr. Rosamma Alex

Preface

Natural rubber is harvested through a simple tapping process from the tree *Hevea brasiliensis*, which is cultivated extensively in tropical and subtropical regions. As a plant-derived polymer, natural rubber exhibits unique properties after processing and vulcanization, including exceptional elasticity, low hysteresis, high strength, and elongation characteristics not found in most synthetic rubbers. Its distinct viscoelastic behavior allows it to behave as a glassy material at high-frequency deformations and as a rubbery material at very low frequencies, even at moderately low temperatures. The load-deformation characteristics of natural rubber are particularly unique. For instance, while rubber is nearly incompressible, its compression modulus can be significantly enhanced when bonded to metal surfaces, all while preserving its inherently low shear modulus.

These remarkable properties make natural rubber indispensable in a wide range of applications, from general mechanical rubber goods to specialized products such as tires, shock absorbers, vibration isolators, and latex-based items like gloves, tubes, sheets, balloons, and catheters. However, achieving these properties requires careful processing operations, including mixing, shaping, vulcanization, and post-vulcanization treatments, all of which are highly energy-intensive. Similar to other chemical industries, the rubber industry also faces environmental and pollution challenges due to the variety of chemicals and processing methods involved.

While significant advancements have been made in the tire sector, processing challenges remain, particularly regarding energy consumption, environmental impact, and health concerns. The rubber industry utilizes a wide range of chemicals, with most being added in small quantities (typically less than 10 parts per 100 parts of rubber). However, fillers such as carbon black and silica are often used in much larger proportions, sometimes exceeding 100 parts per 100 parts of rubber. These particulate fillers, particularly in the tire sector, are often nano-sized, making their incorporation into the rubber matrix both energy-intensive and a source of environmental pollution.

To address these challenges, a more environmentally friendly approach has been explored, incorporating carbon black or silica into rubber through the latex stage, followed by coagulation and drying to produce a filler rubber masterbatch. This book presents the results of recently developed natural rubber–carbon black masterbatches, focusing on the production and processing techniques, flow and vulcanization behavior, prediction of dynamic properties for tire applications, and design aspects of natural rubber–carbon black systems in dynamic applications.

Additionally, we provide detailed mathematical derivations related to flow characteristics of rubber, vibration isolation and transmissibility, and shock transmissibility. The frequency–temperature superposition technique for predicting the dynamic properties of rubber products is also covered. The book also explores compound design and mechanical parameter optimization for various applications, including vibration isolators, bridge bearing pads, and flex seals, with a particular focus on latex-based carbon black masterbatches and latex compounds.

With a balanced blend of theoretical insights and practical applications, this book serves as a valuable resource for students, trainees, researchers, entrepreneurs, and professionals in the rubber industry.

This work is primarily based on my research experience spanning 25 years at the Rubber Research Institute of India, along with the training and guidance I received from highly reputed institutions in India and abroad. I extend my sincere appreciation to my co-authors for their dedicated efforts in bringing this book to fruition. I am also grateful to Hanser Publishers, Dr. Mark Smith, and his team for their invaluable review comments and for agreeing to publish this book.

Foreword 1

Natural rubber was known to be in use for several years by the diverse civilizations in the historical area covering the southern part of North America and coastal regions of Central America. The rubber-producing tree, *Hevea braziliensis*, was one of the native trees in the rainforests of the Amazon region in South America. The first use of the milky exudate from the rubber tree as play balls by the inhabitants of this region was observed by Christopher Columbus and his crew during his voyages. The play balls were taken to Spain in 1496, and presented to the queen of Spain; this was the beginning of the entry of rubber to Europe. The rubber plantation started after the transportation of seeds of the trees collected from the Amazon valley to Kew Gardens, London. From here, the planting materials were transported to other geographically suitable rubber-growing regions and this marked the beginning of rubber plantation in Southeast Asia. The name "rubber" came into use by Joseph Priestly in 1770, because this material could remove pencil marks by rubbing. In French the name was "caoutchouc", derived from the native Amazonian expression "caa-o-chu", meaning "weeping tree", and still remains in common use within the French language.

Two earlier products from rubber were shoe soles and waterproofed fabrics. This is based on reports by a French voyager, Jean Marie de la Condamine, to the Royal Academy of Sciences in 1751. The first commercial rubber product is considered to be waterproofed clothes, and Charles Macintosh of Scotland in 1824 opened a factory for the same in Manchester. In 1820, Thomas Hancock from England developed the process of mastication of rubber using a machine called the pickle machine. In 1836, Edwin Chaffee of Roxbury, Massachusetts, developed the two-roll mill for mixing additives into rubber.

A revolutionary discovery that made rubber an industrial material was in 1839 by Charles Goodyear (1800–1860), who observed the engineering properties of vulcanized rubber through the addition of sulfur. The discovery of vulcanization was made almost simultaneously by Thomas Hancock (1786–1865) (patented May 21st) and Charles Goodyear (patented June 15th).

In 1844, during the Industrial Revolution, the importance of rubber as an indispensable raw material for the rubber industry emerged.

"Basics of Technology and Engineering with Natural Rubber" provides understanding of and information about the processing and engineering aspects of natural rubber. The important aspects of the flow of natural rubber and rubber compounds with reference to its shaping and processing are described. The important applications based on certain unique properties of natural rubber related to viscoelasticity are also considered.

The book is organized into five chapters covering collection, preservation, and marketable forms of natural rubber; rheology of rubber in rubber processing; viscoelasticity of rubber; basic principles of vibration, shock, and sound isolation based on rubber viscoelasticity; and bridge bearing and general elastomer bearings. Adequate attention is given to the basic technological aspects in designing rubber products related to vibration isolation, shock absorption, and elastomer bearings. The specific properties of natural rubber based on viscoelasticity as a green alternative and complement to synthetic rubber are considered in rubber product design. References and appendixes provided in the book give a better understanding of the technical and engineering aspects related to natural rubber.

There are a large number of applications specific for natural rubber because of its unique viscoelasticity, high strength combined with high elongation and elasticity, rubbery characteristics related to vibration, shock, and sound absorption, and electric insulation. Natural rubber exhibits a unique property of strain-induced crystallization and due to this, it exhibits high tensile strength without incorporation of any fillers. The excellent cut growth resistance is also attributed to the strain-induced crystallization of natural rubber. Natural rubber is available as a latex and hence has versatility in the production of a large number of latex-based products. Being available as a latex also has the advantage of environmentally friendly and energy-saving processing operations through latex-stage incorporation of additives and reinforcing fillers.

Although the book provides basic aspects related to natural rubber, the contents are useful in designing general rubber products and new applications of rubber.

It gives me pleasure to introduce this book to all those who are interested in rubber as an industrial raw material, specifically natural rubber farmers, estate managers, technologists, rubber product manufacturers, researchers, and students.

Kottayam, June 2025

Dr. Sawar Dhanania

Foreword 2

Natural rubber, a remarkable biomaterial, has played a pivotal role in shaping the modern rubber industry. While much of the global focus has historically been on tire-related applications, the engineering and design of non-tire rubber products represent a vast and equally significant domain. This book, "Basics of Engineering and Technology with Natural Rubber" focuses on the unique challenges and solutions associated with the compound design processing and manufacture of such products.

The book has five well-structured chapters that offer both theoretical depth and practical insights into technological and engineering aspects related to natural rubber. It is especially valuable in its coverage of the compound design and manufacturing aspects of non-tire applications, ranging from vibration isolators, bridge bearing pads, and engine mounts to dock fenders and flex seals. The authors have presented the mathematical derivations for the engineering principles in a clear and accessible manner, making this book highly relevant not only for students and academic learners but also for entrepreneurs and professionals seeking to understand product design from the grassroots level.

One of the notable contributions of this book is its emphasis on the importance of natural rubber due to its unique aspects including specific mechanical and dynamic properties. The discussion on latex carbon black masterbatch technology has novelty as raw rubber is available as latex that coagulates by simple processing methods. The authors highlight its advantages in achieving an environmentally friendly process for incorporation of fillers, especially carbon black and silica, into natural rubber. The filler-incorporated rubber shows superior filler dispersion, and consequently improved product performance and processability, which is an area of growing relevance in sustainable manufacturing, in both tire and non-tire sectors.

From this book the reader will understand the processing technology of rubber in general and the relevance of natural rubber as a sustainable and high-performance material.

I congratulate the authors for compiling such a comprehensive and thoughtful volume and am confident that this book will serve as a valuable reference for all those engaged in rubber engineering, be it in industry, research, or education.

Kottayam, July 2025

M. Vasanthagesan IRS

Executive Director, Rubber Board

Contents

The Authors V

Acknowledgments IX

Preface XI

Foreword 1 XIII

Foreword 2 XV

1   **Natural Rubber Latex Collection,**
      Preservation and Marketable Forms 1

1.1   Introduction 1

    1.1.1   Natural Rubber Latex: Harvesting, Composition
           and Significance of the Various Latex Components 2

    1.1.2   Significance of Non-Rubber Ingredients 6

        1.1.2.1   Significance of Proteins in Latex and Removal
                 of Leachable Proteins 6

    1.1.3   Factors Affecting Colloidal Stability of Latex 8

        1.1.3.1   Mechanism of Latex Coagulation: Colloidal
                 Destabilization, Coagulation, Gelation and Flocculation   9

        1.1.3.2   Coagulation of Latex by Externally Added Chemicals ... 10

        1.1.3.3   Coagulation without the Addition of External Chemicals   11

        1.1.3.4   Modified Coagulation of Fresh Natural Rubber Latex
                 and Production of Latex Filler Masterbatch 12

1.1.3.5 Latex-Stage Incorporation of Ingredients into Latex and Pre-Vulcanized Latex 14

1.1.3.6 Some Aspects Related to Enhanced Elasticity and Low Hardness in Latex-Based Products 15

1.2 Marketable Forms of Rubber 16

1.2.1 Preserved Latex and Latex Concentrate 17

1.2.1.1 Preservation of Field Latex 17

1.2.1.2 Methods of Latex Concentration 18

1.2.2 Sheet Rubber: Ribbed Smoked Sheet and Air-Dried Sheets 27

1.2.2.1 Sieving and Dilution of Latex 28

1.2.2.2 Estimation of Dry Rubber Content (DRC) of Latex 30

1.2.2.3 Latex Coagulation and Production of Sheets 31

1.2.2.4 Defects in Rubber and Grading of Sheets 32

1.2.3 Block Rubber or Technically Specified Rubber (TSR) 34

1.2.4 Crepe Rubber 39

1.2.4.1 Pale Latex Crepe 40

1.2.4.2 Field Coagulum Crepe 41

1.2.4.3 Reclaimed Rubber 41

1.2.4.4 Retreading of Tires and Reuse of Waste Tires 42

2 Rheology in Rubber Processing 49

2.1 Introduction 49

2.1.1 Deformation of Ideal Viscous Fluids and Ideal Elastic Solids 50

2.1.2 Stress and Strain in Solids 51

2.1.3 Newtonian Flow 52

2.1.4 Non-Newtonian Flow 53

2.1.4.1 Dilatant (Shear-Thickening) Flow 54

2.1.4.2 Bingham Plastic 54

2.1.5 Time-Dependent Viscosity 55

2.1.5.1 Thixotropic Fluids 55

2.1.5.2 Rheopectic Fluids 56

2.2 Rubber as a Viscoelastic Material 56

2.2.1 The Flow of Rubber and Rubber Compounds during Processing to Products 57

2.2.1.1 Elastomer 58

2.2.1.2 Fillers 59

2.2.1.3 Textile Materials in the Rubber Industry 59

2.2.1.4   Processing Aids 61

2.2.1.5   Anti-Degradants 62

2.2.1.6   Vulcanizing Chemicals 62

2.2.1.7   Special-Purpose Additives 62

2.2.2   Rubber Formulation 63

2.3   Processing of Rubber 63

2.3.1   Steps Involved in the Processing of Rubber 63

2.3.1.1   Mastication of Rubber 65

2.3.1.2   Mixing Process on a Two-Roll Mill 65

2.3.2   Influence of Method of Incorporation of Carbon Black Via Latex Stage Incorporation 67

2.3.3   Production of Latex Carbon Black Masterbatch Based on a Modified Coagulation of Natural Rubber Latex – Factory Scale ... 68

2.4   Shaping and Vulcanization 72

2.4.1   Compression Molding 72

2.4.1.1   Mold Shrinkage During Compression Molding 73

2.4.2   Extrusion 73

2.4.3   Calendering 75

2.5   Vulcanization of Extruded and Calendered Profiles 76

2.5.1   Batch Vulcanization Techniques 76

2.5.1.1   Autoclave or Steel Pan 76

2.5.1.2   Lead Curing 77

2.5.2   Continuous Vulcanization Techniques 77

2.5.2.1   Hot-Air Tunnel 77

2.5.2.2   Fluidized Bed 77

2.5.2.3   Liquid-Curing Method (LCM) 77

2.5.2.4   Continuous Drum Cure 78

2.5.2.5   Microwave Vulcanization 79

2.6   Viscoelastic Parameters of Raw Rubber and Characterization 80

2.6.1   Microstructure of NR 80

2.6.2   Strain-Induced Crystallization of Natural Rubber 81

2.6.3   Macrostructure 82

2.7   Viscoelasticity of Raw Rubber and Rubber Compounds 84

2.8   Instruments Used to Measure the Viscoelasticity of Natural Rubber in Relation to Processing 85

2.8.1   Compression Plastimeter: Wallace Rapid Plastimeter 85

2.8.2   Rotational Viscometer: Mooney Viscometer   86

2.8.3   Oscillating Rheometers, Moving-Die Rheometers and RPA 2000 ..   87

2.9   Measurement of Processability as Related to Extrusion   88

2.9.1   Capillary Rheometer   88

2.9.2   Extrudability Performance   91

2.9.3   Extrudability Performance of Rubber Mixes   91

2.9.4   Factors Affecting Extrudate Performance of Dry- and Latex-Stage
         Mixed Mixes   95

         2.9.4.1   Die Parameters Affecting Die Swell   95

         2.9.4.2   Factors Influencing Rubber Flow and Die Swell   95

         2.9.4.3   Effect of Micro- and Macrostructure on Rubber   95

3   Viscoelasticity of Rubber **99**

3.1   Introduction   99

3.2   Brief History of Rubber   101

3.3   Natural Rubber   102

3.4   Vulcanization of Rubber and its Mechanical Properties   102

3.5   Fillers and Mechanical Properties of Rubber Vulcanizates   104

3.5.1   Significance of Carbon Black   106

3.5.2   Significance of Silica   108

3.6   Mechanical Properties Related to the Viscoelasticity of Rubber   110

3.7   Viscoelastic Parameters   110

3.7.1   Stress Relaxation and Creep   111

         3.7.1.1   Stress Relaxation and Permanent Set   112

         3.7.1.2   Compressive Stress Relaxation   115

3.7.2   Hysteresis and Strain Energy   115

3.7.3   Resilience   118

3.7.4   Cyclic Deformation of Rubber: Mullins Effect and Payne Effect
         in Carbon-Black-Filled Systems   119

3.7.5   Phase Lag   120

         3.7.5.1   Dynamic Mechanical Properties of Rubber   120

3.8   Viscoelasticity – Mathematical Models   123

3.8.1   Maxwell Model   123

3.8.2   Voight Model   127

3.9   Influence of Temperature on Viscoelastic Parameters   128

3.9.1   WLF Equation (Williams–Landel–Ferry Equation)   129

3.10   Time–Temperature Equivalence in the Relaxation Process   132

3.11   Dependence of Dynamic Mechanical Properties on Time
       of Deformation or Frequency of Deformation and Temperature   134

       3.11.1   Time–Temperature Equivalence: Significance in the Prediction
                of Properties   136

3.12   Effect of Method of Filler Addition on Dynamic Mechanical Properties ...   139

3.13   Service Life Prediction   141

4   **Basic Principles of Vibration, Shock and Sound Isolation
       Based on Rubber Viscoelasticity**   147

4.1   Introduction   147

4.2   Vibratory Systems   149

4.3   Forced Vibration   150

4.4   Free Vibration   150

       4.4.1   Mathematical Expression for the Natural Frequency
                of a Free-Vibration System   151

       4.4.2   Free Damped Vibration   154

                4.4.2.1   Viscous Damping   154

       4.4.3   Material Damping: Mathematical Equation for Quantifying
                Damping in Terms of Energy Loss   158

4.5   Vibration Isolation   161

       4.5.1   Mathematical Expression for Transmissibility
                in Forced Damped Vibration Isolation   161

4.6   Compound Development for an Engine Mount Based on Natural Rubber   171

4.7   Shock Isolation   174

       4.7.1   Shock Pulse   175

       4.7.2   Mathematical Expression for Velocity Changes Occurring
                During a Shock   176

4.8   Shock Isolation   178

       4.8.1   Mathematical Expression for Linear Deflection of an Isolator
                under Shock   178

       4.8.2   Mathematical Expression for Shock Output (G_{out})   179

       4.8.3   Shock Transmissibility   180

4.9   Rubber as a Sound-Absorbing Material   182

       4.9.1   Propagation of Sound in a Medium   183

       4.9.2   Key Terms Related to Sound Absorption   183

                4.9.2.1   Sound Pressure   183

	4.9.2.2	Sound Power	184
	4.9.2.3	Sound Intensity	184
4.9.3	Sound Absorbers		184
4.9.4	Sound Absorption Characteristics		184
4.9.5	Sound Absorption by Porous Materials		185
	4.9.5.1	Cellular Rubber-Based Acoustic Materials	185
	4.9.5.2	Latex Foam	186
	4.9.5.3	Sponge Rubber	187
	4.9.5.4	Polyurethane Foams	189

5   Bridge Bearings and General Elastomer Bearings   195

5.1	Bridge-Bearing Pads		195
	5.1.1.1	Force Deflection Characteristics of Rubber	197
	5.1.1.2	Young's Modulus (E) or Elastic Modulus	197
	5.1.1.3	Compressive Stiffness	197
	5.1.1.4	Shear Modulus	200
	5.1.1.5	Bulk Modulus (K_b)	201
	5.1.1.6	Poisson's Ratio	202
	5.1.1.7	Stress-Strain Behavior of Rubber	203
5.2	Elastomers Used in Bridge-Bearing Pads		203
	5.2.1	Salient Features of CR	204
5.3	Design of an Elastomeric Bearing: Bridge Bearing		205
5.4	Rubber-to-Metal Bonding		208
5.5	Base Isolation		210
5.6	Rail Pads		214
5.7	Flex Seal Used in Space Applications		215
5.8	Dock Fender		216
5.9	Spheriblock		217
5.10	Rubberized Road		217
5.11	Rubber as Insulating Material		218
5.12	Anti-Static Gloves		220

Index   225

1

Natural Rubber Latex Collection, Preservation and Marketable Forms

Rosamma Alex, Umasankar G, K. Madhusoodanan

1.1 Introduction

Natural rubber has been in use for a very long time in the history of mankind and its use can be traced back to the Maya and Inca civilizations. Due to possibly rubber's elastic and sticky nature, it was used as balls for playing games and religious rituals. Rubber was introduced as a product to Europe for the first time in 1496 by Christopher Columbus, who had observed the inhabitants of the Caribbean islands playing with rubber balls during his voyage. He later brought these balls to Spain. Rubber plantation started after Sir Henry Alexander Wickham, a British explorer, brought rubber seeds from the plant *Hevea brasiliensis* from Brazil to the Royal Botanical Gardens, Kew, London, in 1876 [1]. The seedlings and other planting materials were later sent to the Asian colonies, where large-scale plantations and rubber estates were established on account of the favorable climatic and geographical conditions, especially in Malaysia, Ceylon and Singapore.

Several plants produce natural rubber in the form of latex. *Hevea brasiliensis* is the most important source of NR. However, two plants that have been developed to a limited extent as alternative rubber-producing crops are the guayule shrub (*Parthenium argentatum*) and the Russian dandelion (*Taraxacum koksaghyz*). Russian dandelion has laticifers in the roots, while the guayule shrub is a non-laticiferous rubber-producing plant. The rubber obtained from these plants contains more non-rubber ingredients and has long and difficult processing methods. The rubber yield is comparatively low and rubber is extracted from the roots of Russian dandelion and the roots and other parts of the plants for guayule shrub. Plants other than Russian dandelion and guayule shrub are of only minor significance in respect of quantity, quality, method of latex collection, cost aspects and research. In many species, such as fig trees, the rubber content is too low to be considered a source of natural rubber.

Rubber plantations are very prominent in areas that have an equatorial monsoon climate. Rubber plantations require good rainfall (over 2000 mm annually) temperature of 29–34 °C, moderate sunshine (2000 hours per annum) and relative humidity close to 80% with moderate wind. Rubber plantations are present in southeast Asian countries, including Thailand, Indonesia, Vietnam India, Malaysia, Sri Lanka, Myanmar and Cambodia, in west Africa, the Ivory Coast, Guatemala, Mexico, China and the Philippines. In the biological process of production of latex by the rubber tree, there is sequestering of carbon dioxide from the atmosphere, as carbon dioxide is used by the tree during photosynthesis to produce glucose and release oxygen to the atmosphere leading to the production of biomass, including rubber hydrocarbon.

Natural rubber is a sustainable viscoelastic polymer that forms the primary raw material for various rubber products, ranging from a simple rubber band to aircraft tires. The polymer molecules are produced in the rubber tree *Hevea brasiliensis* by a biosynthesis reaction and harvested in the form of latex. Biosynthesis happens through biochemical reactions, starting with an allylic diphosphate, such as farnesyl pyrophosphate, under suitable biochemical conditions inside the rubber tree.

The production of many products involves converting latex to a solid form, either as raw rubber or a latex-based product. In this context, latex composition, its coagulation behavior, retention of specific non-rubber ingredients on solidified rubber and conversion of fresh field latex into specific modified and marketable forms gain importance. There is a vast variation in the size of rubber particles and the type of proteins and lipids on different rubber particles. The coagulation characteristics of latex and the microstructure of rubber recovered from latex vary for latex-containing rubber particles of comparatively bigger particle size and latex-containing rubber particles of relatively smaller size. Generally, latex obtained by a concentration process separates the latex into a fraction having comparatively small rubber particles (SRPs) and a fraction having large rubber particles (LRPs).

The coagulation characteristics can be utilized to add additives, such as fillers, in the latex stage. The advantage of adding fillers in the latex stage is better filler dispersion and less environmental pollution.

1.1.1 Natural Rubber Latex: Harvesting, Composition and Significance of the Various Latex Components

The rubber tree synthesizes latex in special cells called latex vessels, also known as laticifers, and they appear as concentric layers inside the bark of the *Hevea brasiliensis* tree. The latex is harvested from the tree by a process of tapping. The latex vessels are oriented at an angle of 2–7° from the vertical direction. Removing the bark as thin shavings from high left to low right [2] ensures that the maximum number of latex vessels are cut open. Usually, a half-spiral cut is made at an angle of about 30° to the horizontal direction to allow latex flow. When thin strips of bark very close to the cambium (about 1 mm thick) are removed, the latex vessels are cut open and latex flows out.

After a specific time following tapping, the lactiferous vessels become blocked, due to the action of specific proteins known as lutoids that are found within organelles of latex. The plugging of latex vessels can be a defensive mechanism of the plant against invasion by pathogens, which happens through latex coagulation. The main component of lutoids is a lectin-like protein called hevein. The lutoids rupture under physical shear or other physiological factors, releasing the hevein protein along with divalent calcium ions [3]. It is assumed that, in the presence of hevein, multi-valent linkages are formed between the rubber particles under the conditions existing in the latex vessels, leading to slow latex coagulation [4]. As a result, severed latex vessels are closed by the coagulated rubber and latex ceases to flow.

The ability of the tree to plug the latex vessels by coagulating rubber varies with different clones. In trees subjected to intensive tapping frequency, latex can be pre-coagulated on the tapping panel and the latex flow can be reduced. Sometimes, there can be late dripping of rubber latex. This is due to faster lysis of lutoids, leading to latex coagulation; this condition is referred to as tapping panel dryness [5].

Tapping is done in the early hours of the day because the turgor pressure in the latex vessels (the hydrostatic pressure generated by the latex itself) is highest at that time, at 10 to 15 atmospheres.

The latex harvested is called field latex and is subjected to preservation and concentration to produce commercial rubber products.

Latex can be considered as the cytoplasm of laticifer cells, and so, in addition to its main component, the rubber hydrocarbon, it also contains many living components of plant cells and biochemical compounds. These various latex components have a unique physiological role and possess significant technological attributes that affect the properties of latex and dry rubber.

The ultracentrifugation process can separate latex components into three principal fractions [6, 7]. The rubber is present in the form of cream as the top fraction. Another fraction just below the cream fraction appears as a yellow layer and consists of Frey-Wyssling particles. The next fraction is the cytoplasmic serum (C-serum), which contains mainly soluble components, such as proteins, carbohydrates, organic acids and inorganic salts. The bottom fraction consists of organelle-like particles called lutoids. The aqueous serum in the lutoid fraction is obtained as a bottom fraction called B-serum.

Other than rubber particles, two other specialized structures found in the cytoplasm of laticifer cells of the *Hevea brasiliensis* tree in the form of particles are lutoids and Frey-Wyssling particles. The yellowish color of dry rubber stems from carotenoid-type pigments in Frey Wyssling particles [8]. Fresh latex has a specific gravity of 0.96–0.98 and a pH of 6.5–7.0. The pH of latex changes slowly when exposed to the atmosphere. The significance of specific gravity is that this parameter can be correlated with the dry rubber content of latex. In addition to rubber and water, the main biochemical compounds in latex are proteins, lipids (phospholipids, glycolipids, fatty acids), resinous materials (sterol and terpenoids), carbohydrates (quebrachitol, sucrose) and inorganic ions (K^+, Mg^{2+}, Cu^{2+}, $Fe^{2/3+}$, Na^+, Ca^{2+} and PO_4^{3-}). The composition of fresh latex is rubber hydrocarbon (28–44%), proteins (1–1.6%), resins (1.5–2.5%), mineral matter (0.7–0.9%), carbohydrates (0.8–1%) and water (55–60%). The main carbohydrates are cyclitols, such as L-quebrachitol, sucrose and glucose. An appreciable quantity of non-rubber constituents remains adsorbed on rubber particles or is associated with rubber hydrocarbon, and the remaining parts are dispersed in the serum. Natural rubber latex serum contains several chemicals that can have industrial applications. L-quebrachitol is believed to possess medicinal properties and some of the proteins may have potential for industrial application [9].

Since there are non-rubber constituents, mainly proteins, amino acids and phospholipids, associated with rubber particles, the rubber recovered from latex contains some non-rubber ingredients.

As the main component of natural rubber latex is rubber particles, latex can be considered as a colloidal dispersion of rubber particles in an aqueous medium. The rubber content of fresh, natural rubber latex depends on the clone, the climatic conditions and the plantation management techniques followed. Dry rubber content (DRC) is the mass in grams of rubber present in 100 g of latex recovered by slow coagulation under suitable conditions of dilution and as per national standards. DRC is important, as transactions and chemical additions are made on the basis of this parameter. DRC is also used to assess the quality of latex [15].

Rubber particles in fresh natural rubber latex are generally spherical, but can also be pear-shaped, with sizes varying from 0.01–5 μm, the majority being 0.1–2 μm. The particle distribution is suggested as bimodal [10], constituting mainly large rubber particles (LRPs) and small rubber particles (SRPs). The average particle diameter of SRPs is 10–250 nm and that of LRPs is 250–3000 nm [11].

A study shows that the particle size of preserved concentrated latex varies from 480–600 nm [12].

A rubber particle that contains many rubber molecules as a central core is envisaged as being bound to an outer layer of proteins through an inner layer of phospholipids [11]. Specifically, there can be mixed layers of proteins and phospholipids, while the surface of freshly tapped latex differs from that of preserved latex concentrate [13]. The rubber molecules present in the rubber particle are attached to specific groups at the two ends. The initiating group of a rubber molecule is referred to as ω-terminal and the group at the other end, which is the terminal group, is called α-terminal. A rubber molecule contains an initiating group followed by 0–3 trans-1,4-isoprene units, a long sequence of cis-isoprene units (1000–3000) and a terminal group [14].

The α-terminal group of the NR molecule is considered to contain phosphate functional groups that may be either monophosphate or diphosphate [15, 16, 17]. These groups facilitate the linkage of phospholipids to rubber molecules through hydrogen bonding and ionic bonding [13]. The ω-terminal interacts with protein molecules consisting of amino acids. The protein molecules appear as if they extend from the surface of rubber particles into the latex serum [15]. The proteins are susceptible to attack by bacteria, which leads to latex putrefaction, and are also likely to be displaced by fatty acid soap anions formed naturally by the hydrolysis of phospholipids of latex [18]. A pictorial representation of a rubber particle in freshly harvested natural rubber latex before spontaneous coagulation is shown in Figure 1.1 [13].

The carboxyl groups of the amino acids in the protein adsorbed on rubber particles become partially dissociated, providing a negative charge on the outer surface of rubber particles. Proteins in the outer layer, which can interact with phospholipids to form a protein-phospholipid layer, provide colloidal stability to *Hevea* latex. A second role of phospholipids in enhancing colloidal stability is that they slowly hydrolyze to give fatty acids, which ionize, become adsorbed onto the rubber particle surface and provide a barrier against rubber particle aggregation. The phosphate groups also provide a negative charge for the rubber particles. Repulsion among the like-charged rubber particles imparts colloidal stability to latex.

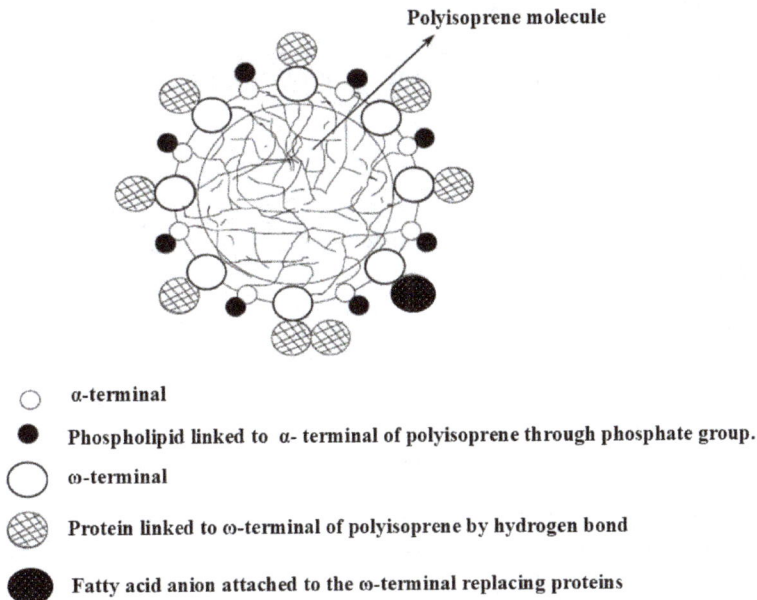

Polyisoprene molecule

○   α-terminal

●   Phospholipid linked to α- terminal of polyisoprene through phosphate group.

◯   ω-terminal

▨   Protein linked to ω-terminal of polyisoprene by hydrogen bond

⬤   Fatty acid anion attached to the ω-terminal replacing proteins

Figure 1.1 Pictorial representation of a rubber particle in a freshly harvested natural rubber latex before spontaneous coagulation. Modified and redrawn with permission [13]

Natural rubber is a polymer of cis-1,4-polyisoprene (contains more than 99.9% of cis-1,4 structural units). It has a very high molecular weight, which mainly varies with the clones and the age of rubber trees, with a small contribution from management practices along with the soil and climate conditions of the plantation. The molecular weight (weight average molecular weight M_w) of the soluble portion of natural rubber exceeds 1×10^6 Daltons. It can be close to 1×10^7 Daltons, with a wide range of molecular weight distribution (MWD) and a polydispersity index ranging from 2 to 11 [17]. The molecular weight follows a bimodal distribution [19]. Natural rubber's processability and physical properties are closely related to its molecular weight and distribution.

1.1.2 Significance of Non-Rubber Ingredients

Non-isoprene constituents, such as tocotrienols, proteinsand amino acids, function as weak anti-oxidants, while fatty acids can act as vulcanization activators. Some inorganic constituents, such as copper, manganese and iron are deleterious pro-oxidants, because they accelerate thermal degradation associated with rubber oxidation.

1.1.2.1 Significance of Proteins in Latex and Removal of Leachable Proteins

One of the significant biochemical components of natural rubber latex is proteins (1–1.5%). They are found distributed on the rubber particles (25–27% of total proteins), latex serum (43–48% of total proteins) and lutoids (25–32% of total proteins) [20].

Proteins play a significant role in latex processing:

a) The colloidal stability of latex is attributed to many types of proteins adsorbed on rubber particles. The adsorbed proteins are acidic and insoluble in aqueous medium. The isoelectric point of these proteins adsorbed on the rubber particle surface is close to 4.7.

b) Proteins are amines and affect the technological and processing aspects of rubber. They have an accelerating effect on the vulcanization of rubber and act as anti-oxidants, as evidenced by the plasticity retention index (PRI) measurement. However, the dynamic mechanical properties, such as heat build-up, compression set and hardness, are adversely affected by proteins.

c) Many proteins from NRL, including those adsorbed-on rubber particles in aqueous serum and those in lutoids, cause allergies such as skin irritation and other general allergy problems, including fatal anaphylactic reactions. At least ten to thirteen latex proteins are allergens [21]. Latex allergy is caused mainly by water-soluble proteins that come into contact with our bodies. The two main allergens identified from proteins enveloping the rubber particle are Hev b 1 and Hev b 3. The proteins on rubber particles can change to low-molecular fractions and hence become soluble and allergenic. C-serum contains a large number of proteins that are mostly acidic. Hev b 5, Hev b7, hev b 8 and Hev b 9 are the main allergens in C-serum found in latex serum. The main protein released by the rupture of lutoids is hevein, which is anionic and possibly composed of sulfur-containing amino acids, such as cysteine, methionine and other amino acids and is related to allergic reactions. Hev 2, Hev 4, Hev 6 and Hev 10 are bottom-fraction allergens [22, 23].

Treatment of NRL with preservatives such as ammonia leads to protein hydrolysis. The process of pre-vulcanization by sulfur and accelerators increases the formation of low-molecular water-soluble proteins. Most proteins are soluble, while a few associated with latex organelles are insoluble. The use of latex products, such as rubber teats, condoms, catheters and gloves, which are intended for body contact, can cause allergies when proteins that are leachable by body fluids are present in them.

The leachable proteins in products can be reduced in two main ways:

- Through the use of low-protein latex and

- Through the removal of proteins from the product (gloves, catheters and the like)

Low-protein latex is mainly produced by the degradation of proteins using enzymes and exposure to gamma radiation [24, 25], followed by concentration in the form of either creaming or centrifuging [26, 27]. The polypeptide chain is cleaved by proteolytic enzymes such as papain, present in papaya plant latex, or by alcalase, an enzyme obtained from *Bacillus licheniformis* bacteria [28, 29]. Double centrifugation also results in low-protein latex. The colloidal stability of latex reduces when the proteins are removed, so low-protein latex has a lower colloidal stability. The removal of proteins adversely affects rubber's anti-oxidant properties, as rubber recovered from deproteinized latex has a low PRI and tends to dry with increased tackiness at comparatively higher temperatures.

Key techniques for removing proteins from gloves are leaching, chlorination and surface coating. The leaching process exploits the tendency of proteins to migrate to the glove's surface. So, suitable leaching of gloves in the wet-gel form and the vulcanized form can remove an appreciable quantity of leachable proteins [26]. Chlorination involves a certain degree of protein denaturing and, due to this low molecular weight, protein fragments show less tendency to migrate to the surface. Discoloration and reduction in mechanical properties happen if the chlorination process is not well controlled. A 10% sodium hypochlorite solution mixed with 0.3 g conc. HCl and 97 g water serve as the chlorinating solution. Polymer coating provides a barrier against protein migration. The polymers used can be acrylic polyurethane or silicone [27].

1.1.3 Factors Affecting Colloidal Stability of Latex

Natural rubber latex is a lyophobic colloid and the natural tendency from the thermo-dynamic point is a coalescence of rubber particles. As a colloidal dispersion, the rubber particles exhibit random Brownian motion. Different forces exist in natural rubber particles, including attractive and repulsive forces. A balance of these forces governs colloidal stability. The forces are as given below [30].

(a) Van der Waals forces are attractive forces between rubber particles, similar to a general colloidal system. Van der Waals attractive forces can stem from interactions between all the atoms in one particle and those in a neighboring particle.

(b) Depletion forces: These can be considered as an attractive force that occurs between rubber particles and solutes in aqueous serum. Depletion forces can also be attributed to chemicals, such as ammonium alginate, carboxymethyl cellulose and hydroxymethyl cellulose. These chemicals adsorb onto rubber particles and combine them to form a loose network of rubber particles that slowly separate as cream from the serum.

(c) Electrostatic forces: The repulsive forces are mainly due to the negative charges present on rubber particles. Rubber particles are stabilized by a protein layer that is in ionic interaction with phospholipids and this envelope gives the rubber particles a negative charge that results in repulsive electrostatic forces between rubber particles. There will be counter-ions near each rubber particle, due to the negative charge of the phospholipid-protein layer enveloping rubber particles; the result is an electrical cloud called an electrical double layer. The electrical double layer can offer colloidal stability. Adsorption of suitable anions on rubber particle surfaces can increase the electric potential of the rubber particles.

Increasing the aqueous phase's ionic strength can compress the electrical double layer and reduce colloidal stability. An externally added electrolyte solution can compress this electric cloud, reducing the repulsive electrostatic forces.

As rubber colloidal particles are charged, there is an electric double layer around the rubber particles that has an associated zeta potential. Preservation with ammonia causes hydrolysis of phospholipids and proteins, leading to a lowering of the zeta potential, as shown in Table 1.1. Fresh latex can be protein-stabilized, while centrifuged latex and preserved latex are predominantly fatty-acid-soap-stabilized rather than protein-stabilized with fewer proteins.

The colloidal stability of latex decreases in the presence of salts, due to the compression of the electrical double layer.

(d) Steric forces: Generally, macromolecules adsorbed on rubber particles provide steric stability by preventing particle aggregation. The rubber particles can also adsorb long-chain fatty acids, probably from the hydrolysis of fresh latex phospholipids. This can also provide colloidal stability, due to steric action, and increase the potential on the rubber particle surface.

Table 1.1 Zeta Potential of Fresh and Ammonia-Preserved Natural Rubber Latex

Type of Natural Rubber Latex	Zeta Potential (ζ) at pH 9 (mV)
Fresh field latex	−62
Preserved NR latex	−58
Centrifuged NR latex	−55

(e) Solvation forces: Adsorption of polymeric materials on colloidal particles can create a barrier by not being soluble in general colloidal destabilizers, such as acids. Ethylene oxide condensate, a latex stabilizer, can provide stability by preventing aggregation of rubber particles in the presence of acids, because the stabilizer is non-ionic and does not react with the H^+ ions of acids.

1.1.3.1 Mechanism of Latex Coagulation: Colloidal Destabilization, Coagulation, Gelation and Flocculation

The two key destabilization processes are coagulation and gelation. Coagulation is generally a slow process in which rubber separates from the serum as a wet coagulum. In gelation, the latex initially changes to a gel with no serum separation. However, there may be shrinkage of the gel later and separation of a very small quantity of serum. Flocculation refers to the partial coagulation of latex into small flocs while the rest of the latex remains in a fluid state.

Gelation is used in the manufacture of expanded rubber, such as latex foam. Zinc oxide, in association with ammonium ions and carboxylic acids, can cause thickening of latex, leading to gelation. Zinc oxide alone does not cause thickening. When added to latex, it interacts with ammonia and ammonium ions to form zinc–amine complexes containing one to four amine ligands. Zinc–amine complexes with less than four ligands decompose to form zinc ions. This decomposition is accelerated by heat. The zinc ions react with carboxylic soap and proteins adsorbed on rubber particles, thereby depriving the rubber particles of latex stabilizers. The reaction of zinc ions with carboxylate anions results in the formation of insoluble zinc soaps that lead to the thickening and gelation of latex [12, 31]. Excessive gelation can cause latex to coagulate. Another gelling agent, sodium silicofluoride, is also utilized in the manufacture of foam.

Both electrostatic destabilization through the reduction of repulsive forces and aggregation of rubber particles through a change in the structure of the protein phospholipid complex are required for latex coagulation.

Any agency such as mechanical agitation or the presence of chemicals whether formed in the latex or added to the latex that neutralizes the charge on rubber particles and allows inter-particle aggregation can result in latex coagulation.

1.1.3.2 Coagulation of Latex by Externally Added Chemicals

Marketable forms of rubber, such as ribbed smoked sheets and certain grades of technically specified rubber, are made by coagulating fresh latex.

Coagulation caused by externally added chemicals can be achieved by adding volatile fatty acids, such as formic acid or acetic acid, non-volatile acids, such as sulfamic acid (a crystalline solid soluble in water), dilute sulfuric acid and salts, such as calcium salts. Latex can also be coagulated in the presence of a lower quantity of formic acid along with salts such as calcium chloride and sodium chloride. In the presence of salts, the ionic strength of the serum phase increases and the electrical double layer is compressed. This can lead to enhanced interaction of stabilizers on rubber particles with acids. Divalent metal ions, such as calcium chloride solution, and surfactant can act as a coagulant for latex [32].

The presence of formic acid or acetic acid reduces the pH of latex close to a value of about 4.5–4.7. The rubber particles lose the electrostatic stability in this pH, which is assumed to be the isoelectric point of proteins covering the rubber particles and so they slowly become destabilized by losing the negative charge. Aggregation of rubber particles may occur, due to the interaction of different types of proteins with phospholipids involving different rubber particles.

The two key proteins on the rubber particle surface are the rubber elongation factor (REF) protein (Hev b1) present mainly on large rubber particles and the small rubber particle (SRP) protein, present mainly on small rubber particles of natural rubber latex. Different types of phospholipids are present on rubber particle surfaces; the major type of phospholipid is phosphatidylcholine. Of the two key proteins present on the rubber particle surface, REF protein has a stronger interaction with phospholipids than SRP protein [33]. The phospholipid-protein interaction varies with the type of proteins and type of phospholipids. It can be assumed that protein phospholipid interaction among different rubber particles can also lead to the aggregation of rubber particles. The aggregation of rubber particles can lead to coagulation. The protein-stabilized system undergoes destabilization through neutralization of the charge on rubber particles and through protein-phospholipid interaction among rubber particles that happens slowly and homogeneously.

The coagulation process of fresh natural rubber latex sets in about four hours after addition of acids. Proteins have comparatively slow reactions with acids. The slow coagulation ensures uniform latex coagulation, changing it to a soft coagulum that can be sheeted easily. The slow coagulation of fresh NR latex changes to a faster coagulation after preservation with ammonia.

For this reason, generally, preserved latex cannot be used to prepare rubber sheets. This is due to chemical changes that mainly occur in the proteins and phospholipids present in latex. The main chemical change that happens during the storage of ammonia-preserved concentrated latex is that the phospholipids in the latex hydrolyze to give free fatty acids that react with ammonia to form fatty acid soaps. Proteins also hydrolyze to form amino acids, which reduce the colloidal stability of latex. The fatty acid anions adsorb on rubber particles and increase the colloidal stability of latex. Fatty acid soaps are more reactive with acids, so they become desorbed from the rubber particle surface and react with the acids or cations, which leads to loss of charge on rubber particles and, hence, destabilization of rubber particles.

Given a short duration of storage of ammonia-preserved field latex, up to about a week, coagulation is slow. It can be assumed that the phospholipid hydrolysis level is insufficient to produce fatty acid soap anions that displace proteins on the surface of rubber particles. Good quality rubber sheets can be prepared from 1% ammonia-preserved fresh latex stored for one week by coagulation with a 2% formic acid solution. Such sheets show acceptable values of raw rubber properties, such as PRI and initial plasticity (P_0). Generally, the PRI of rubber recovered from ammonia-preserved latex has an inferior PRI, due to fewer proteins retained on rubber after coagulation. Some proteins are removed during hydrolysis as water-soluble forms.

However, field latex preserved for more than two weeks coagulates quickly upon addition of acids. It is assumed that sufficient protein hydrolysis has occurred so that more proteins become displaced from the surface of rubber particles. In contact with acids, the fatty acid soap anions become desorbed from the rubber particle surface, leading to colloidal destabilization.

Freshly centrifuged latex exhibits relatively low mechanical stability, because some proteins, which contribute to its colloidal stability, are removed during the centrifugation process. However, during storage, the colloidal stability of latex improves, due to the adsorption of higher fatty acid anions formed from the phospholipids in latex onto the surface of the rubber particles.

1.1.3.3 Coagulation without the Addition of External Chemicals

Plugging of latex vessels

The plugging of latex vessels is the result of a coagulation process that happens without addition of acids. When the lutoids burst, the proteins and divalent calcium ions are released into the serum. Heveine is a lectin-like protein. This protein can bind to specific sugar moieties in glycosylated proteins to form a multivalent linkage between rubber particles, leading to the aggregation of rubber particles and a reduction in colloidal stability. The calcium ions released during the bursting of lutoids can increase the ionic strength of serum and affect the electrical double layer surrounding rubber particles. These factors lead to latex coagulation and plugging of latex vessels [34].

Spontaneous coagulation of latex

If latex is just kept open for a few hours (4–10 hours), latex undergoes spontaneous coagulation. This coagulation generally does not go to completion and is accompanied by an irritating smell. Spontaneous coagulation is usually attributed to microbial action happening in latex. Latex becomes contaminated by different bacteria and yeasts soon after tapping. Some bacteria use carbohydrates for their metabolic activities and convert them into volatile fatty acids such as formic and acetic acids. These volatile fatty acids neutralize the negative charge on latex. Many microorganisms attack the proteins in latex, decomposing them to sulfur dioxide and hydrogen sulfide, producing a foul odor which occurs at a later stage and is often called putrefaction. Both these changes lead to the destabilization and coagulation of latex. It is generally observed that the pH of latex does not go to a low value in the range of 4.5–4.7. In fact, it falls to the pH 6.0–6.3 range. Adding fatty acid soaps can speed up spontaneous coagulation and this acceleration of coagulation depends on the presence of calcium and magnesium ions in latex. From these observations, it is believed that the phospholipids in latex undergo hydrolysis to fatty acid anions and become adsorbed on rubber particles after displacing some proteins. These anions interact with divalent cations, such as calcium and magnesium ions in latex, to form insoluble soaps that destabilize the rubber particles.

Coagulation arising from this interaction can be affected by mechanical forces, such as high-speed stirring. It can also lead to aggregation of rubber particles and is assumed to be mainly due to protein phospholipid interaction that leads to aggregation and coagulation [35].

The different types of coagulation affect the processing and technological properties of latex, as the types of non-rubber ingredients retained on rubber are different. The amount of biochemical compounds, such as proteins and lipids, retained on the rubber during coagulation and other chemicals added or formed during coagulation play specific roles in its properties.

Any process that removes the protective layer that affords colloidal stability can also lead to coagulation of latex, as is observed when the latex is exposed to microwave radiation (microwave heating).

1.1.3.4 Modified Coagulation of Fresh Natural Rubber Latex and Production of Latex Filler Masterbatch

The conventional slow coagulation of fresh field latex can be changed to rapid coagulation by adding fatty acid soaps to latex. In the presence of an added fatty acid soap, the fresh field latex changes from a protein-stabilized system to a soap-stabilized system. When latex stabilizers, such as fatty acid soaps, are added to latex, they adsorb onto rubber particles by displacing the proteins. However, being more reactive with acids than proteins, they desorb from rubber particles upon addition of acids and the latex coagulates immediately. The reaction of fatty acid soap anions with acids is evident from the quick coagulation of latex treated with fatty acid soap. This coagulation can be used in the production of latex filler masterbatches. When fillers such as carbon black or silica are added to latex, they become coagulated along with latex, since the coagulation happens immediately.

Preparation of latex-stage-incorporated fillers by modified coagulation involves the following stages [36]: (a) addition of a sufficient quantity of fatty acid soaps to sensitize latex to quick coagulation, (b) dilution of latex to about 20 DRC, (c) coagulation by addition of formic acid, (d) creping, (e) washing and drying coagulum at 70 °C. 20% filler dispersions can be made by conventional methods, such as ball milling. The DRC of latex needs to be estimated to ensure the incorporation of the required quantity of filler in the dry rubber. The dried filler–rubber masterbatch is processed in the same conventional way as dry natural rubber (Figure 1.2).

Figure 1.2 Latex carbon black masterbatch (Courtesy: Ms Asiatic rubbers, Kottayam, Kerala)

Effect of adding fatty acid soaps

Fresh, natural rubber latex does not undergo quick coagulation upon addition of small quantities of fatty acid soap. However, if a higher amount is added, the latex coagulates immediately. This coagulation happens, because the quantity needs to be sufficient to displace the proteins and increase colloidal stability, as may be seen from zeta potential values.

Upon addition of fatty acids, the zeta potential is only marginally affected at low concentrations, but increases slightly at higher concentrations of fatty acid soap. There is no increase in zeta potential at a very low concentration of 0.1% (w/v on latex). However, there is an increase in zeta potential at higher concentrations of fatty acid soaps (1.0% w/v on latex). The colloidal stability of surfactant-treated latex increases as surfactant concentration increases. The effect of surfactant addition on the zeta potential of fresh, natural rubber latex is shown in Figure 1.3. Zeta potential of fresh latex and fatty acid soap-treated fresh latex from pH 4–12.

The zeta potential of fresh latex and fatty acid soap-sensitized fresh latex was determined with the aid of a Malvern Zetasizer, model Nano-S (U.K.). The latex samples had surfactant concentrations varying from 0.1–1.0% (w/v on latex). 10 mL of fresh, natural rubber latex of 33% DRC is mixed with different concentrations of 10% fatty acid soap from 0.1 mL to 1.0 mL to yield concentrations ranging from 0.1–1.0% (w/v) fatty acid soap in latex. The zeta potential is measured by adding one drop of surfactant (fatty acid soap)-treated latex to 5 mL of buffer solutions with pH levels ranging from 4–12.

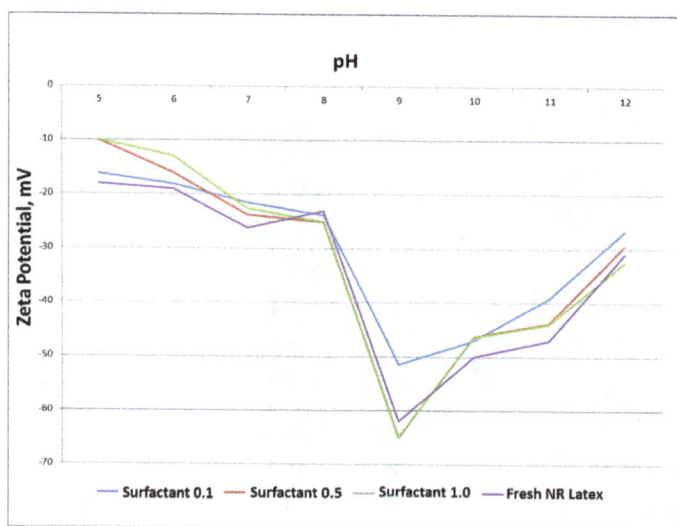

Figure 1.3 Zeta potential of fresh latex and fatty acid soap-treated fresh latex from pH 4–12

1.1.3.5 Latex-Stage Incorporation of Ingredients into Latex and Pre-Vulcanized Latex

Mechanical properties can be imparted to latex by incorporating ingredients such as vulcanizing chemicals and other additives. Pre-vulcanized latex is formed when the rubber molecules inside the rubber particles are cross-linked in the presence of vulcanizing ingredients, such as dispersions of sulfur, accelerator and activator in the latex stage. Suitably compounded latex is heated and maintained at a temperature of 55 ±1 °C using a water bath for a specific time, ranging from one to four hours, to allow the cross-linking of rubber molecules as the vulcanizing chemicals diffuse into the rubber particles. If compounded latex is heated for four hours, the latex obtained is considered a fully pre-vulcanized latex. The chloroform number test determines the vulcanization level of pre-vulcanized latex.

A certain amount of pre-vulcanized latex is mixed and stirred with an equal amount of chloroform. The appearance of coagulated mass can be soft, tacky mass or crumbled particles. Fully pre-vulcanized latex coagulates as a crumbled mass and is assigned a chloroform number 4, while soft tacky coagulum is assigned a chloroform number 1. The coagulum of a sample with chloroform number 2 is a soft, tacky mass, but breaks on stretching while coagulum with chloroform number 3 is not tacky and breaks on stretching.

In products such as gloves, thread and foam, a maturation stage is necessary after addition of chemicals. During latex maturation, vulcanizing chemicals interact with rubber, resulting in the cross-linking of rubber molecules.

Pre-vulcanization can be either full or partial, depending on the product requirements. Fully pre-vulcanized latex is used for products such as balloons, where no further vulcanization is needed. Natural rubber is available as latex and hence has the versatility in processing operations to produce many latex-based products. The availability as a latex also offers the advantage of environmentally friendly and energy-saving processing operations through the latex-stage incorporation of additives and reinforcing fillers.

1.1.3.6 Some Aspects Related to Enhanced Elasticity and Low Hardness in Latex-Based Products

Enhanced elasticity and softness in latex products are achieved by optimizing vulcanization of both the type of cross-links and the density of cross-linking, the quality of the centrifuged latex, the removal of non-rubber ingredients and optimization of the particle size distribution of latex.

This is achieved by:

- Vulcanizing at low temperatures and using a suitable level of latex pre-vulcanization for making latex-based products
- Adopting suitable leaching operation to ensure good film formation and removal of non-rubber ingredients
- Using double-centrifuged latex to ensure the removal of non-rubber ingredients, especially proteins
- Adopting a suitable storage period for preserved latex concentrate before producing latex products
- Adopting any suitable deproteinization method and using deproteinized latex
- Using latex concentrate with bigger particle size
- Ensuring a good level of vulcanization using sulfur and an accelerator system, as employed in conventional vulcanization, and possibly ensuring more disulfidic cross-links and vulcanization at comparatively lower temperatures

The elasticity can be fully evident in latex-based products that are vulcanized at lower temperatures. In latex-stage processing, the chance of molecular-weight reduction is very low, due to the absence of mastication, and mixing on mixing equipment and vulcanization at comparatively higher temperatures. Suitably compounded gum natural rubber-based sheets from pre-vulcanized latex find extensive application as protective linings in coal mines on account of their highly elastic, abrasion resistant nature and low hardness.

1.2 Marketable Forms of Rubber

Rubber productivity in plantations varies with clones and other geographical and climatic conditions. It is usually 50–60 g per day from a mature rubber tree (for Clones RRII 105 and RRII 400 series).

Some 80–88% of the crop is harvested as latex and the remaining as field coagulum. The field coagulum is the dried latex on the tapping panel (tree lace) and the collection cups (shell scrap and cup lumps). The latex and field coagula harvested from rubber plantations are highly susceptible to bacterial contamination, which leads to spontaneous latex coagulation and to a foul smell induced by putrefaction. The decay of proteins by microorganisms results in the generation of foul smells from the degradation products. For these reasons, latex and field coagula must be processed into suitable dried or preserved forms that allow safe storage and marketing [5].

Spontaneous coagulation and latex putrefaction can be controlled with chemicals that suppress bacterial growth, generally called preservatives. Chemicals used for short-term preservation of latex are called anti-coagulants. Chemicals used for long-term preservation need to maintain the quality parameters of latex during storage over a more extended period.

The commonly used anti-coagulants are sodium sulfite, ammonia and formalin. Ammonia and formalin are not utilized as anti-coagulants in the manufacture pale latex crepe. All three anti-coagulants can be used in the manufacture of rubbed smoked sheets.

The four marketable forms of natural rubber are (1) ribbed sheets, (2) technically specified rubbers (TSRs) (3) crepe rubber (pale latex crepe and estate brown crepe) and (4) preserved field latex and latex concentrate. The choice of processing method for the different marketable forms depends on factors such as the availability of crop investment capacity, technical human resources and market demand. A statistical report showed that India had a natural rubber (NR) production of 775,000 metric tons from 2021 to 2022 [37]. During this period, the conversion of the crop to marketable forms was 63.4% ribbed smoked sheet (RSS), 20.0% technically specified rubber (TSR) and 14.1% latex concentrates. A higher percentage of the crop is processed as ribbed sheets, mainly because of the simple and cheap processing techniques involved.

1.2.1 Preserved Latex and Latex Concentrate

Field latex requires preservation and concentration before its commercial use and production of latex-based products. The preservatives added to latex suppress spontaneous coagulation and putrefaction and improve the latex's stability for long-term storage. The concentration of latex involves the removal of a considerable quantity of water so that the dry rubber content of latex rises to a higher value that may be in the range of 55–60%, which is the recommended value as per standard specifications. The transporting of concentrated latex is therefore more economical than transporting field latex. Concentrated latex is the raw material for many latex-based products. The latex concentrate has the advantage of being more uniform in quality than field latex, due to blending field latex from different sources before the concentration process.

1.2.1.1 Preservation of Field Latex

The ideal preservative needs to enhance the colloidal stability of latex by increasing the charge on the particles and the zeta potential at the rubber-water interface. The isoelectric point of proteins, which provide colloidal stability to latex, falls within the acidic range. Maintaining the pH of latex in alkaline conditions is significant, as the negatively charged hydroxide ions enhance the negative charge on rubber particles. Therefore, the preservative needs to be an alkali. It should either be a bactericide or suppress or inactivate bacterial action. The preservative should also deactivate trace metals, particularly metal ions of copper, manganese and iron, by sequestration or precipitation. Removal of metal ions is desirable for two reasons. (1) Some of these metal ions present in latex are essential for the well-being of microorganisms and provide congenial conditions for the growth of microorganisms. (2) Some metal ions reduce the colloidal stability of latex by forming insoluble soaps and affect the quality of rubber by acting as pro-oxidants. In addition, a preservative should not be harmful to people, should not enter into adverse reactions with rubber or containers of latex and should be cheap, readily available and convenient to handle.

Ammonia is an acceptable preservative for latex. At low ammoniation levels (0.05% m/m), the pH of latex increases to about 8, which is conducive to the proliferation of ammonia-resistant bacteria. However, ammonia is a bactericide at higher concentrations (above 0.35% m/m). Ammonia improves colloidal stability. Ammonia also helps to remove metal ions that affect the colloidal stability and properties of latex. For instance, Mg^{2+} ions in latex reduce latex's colloidal stability or create problems during processing by forming insoluble soaps or producing a sludge that may settle at the bottom of the storage tank. Ammonia can precipitate the Mg^{2+} ions present in latex as magnesium ammonium phosphate. Ammonia deactivates copper ions – which are pro-oxidants for the aging of rubber – through complex formation.

Typically, ammonia is used at a concentration of 1% in latex. For producing preserved field latex, the latex is passed through a 40-mesh sieve to remove bark, leaf or other particles, homogenized in large tanks and purged with ammonia gas to become a minimum concentration of 1% w/w (1 kg of ammonia in the form of gas is required for 100 kg of latex). Field latex is marketed in coated/painted mild steel barrels of 200 L capacity.

Ammonia has an obnoxious smell and may irritate skin. For the production of many products, ammonia content needs to be lowered. Hence, low-ammonia preservation systems consisting of an alkali, colloidal stabilizer and a bactericide have become popular. A commonly used system (called the low ammonia TMTD-ZnO system (LATZ) system) contains 0.3% ammonia, 0.0125% tetramethylthiuram disulfide (TMTD), 0.0125% ZnO and 0.05% lauric acid.

Ammonia, commercially available as a 25% solution, can also be used for preservation. Both high and low ammonia preservation is used for field latex.

The chemicals used as latex preservatives exhibit toxicity problems at different levels. Ammonia-free preservatives can be based on a combination of a non-toxic bactericide and a latex stabilizer. Studies reveal the use of dodecyl benzenesulfonic acid (DBS) in the ammonia-free preservation of latex [34]. However, the availability of commercial ammonia-free safe preservatives is limited.

1.2.1.2 Methods of Latex Concentration

The four methods used for latex concentration are (1) creaming, (2) centrifuging, (3) electrodecantation and (4) evaporation. Evaporation removes only water; hence, the total solids content (based on the original latex) and the particle size distribution remain unchanged. However, the other methods partially remove non-rubber constituents and separate latex into fractions containing smaller and larger rubber particles. Centrifuging and creaming are more commonly employed methods for producing concentrated latex, with electrodecantation used only on rare occasions [38].

Creaming

Creaming is a process involving the addition of creaming agents to latex and allowing separation of latex as creamed and serum fractions (Figure 1.4). There is a density difference between rubber and aqueous serum. (The specific gravity of rubber is 0.92, while that of aqueous serum is close to 1.02.) Rubber particles have a natural tendency to rise to the top, as their density is lower than the aqueous serum of latex.

The mechanism of the creaming process is generally considered obscure. However, creaming in latex can be attributed to the ability of macromolecules, which exist as colloidal particles, to adsorb surface-active materials.

Creaming agents, such as alginates and fatty acid soaps, are adsorbed to rubber particles. These adsorbed ions can interact such that the rubber particles form a loose network. Being lighter than the serum, the loose network between the molecules of the creaming agent adsorbed upon the surface of the rubber particles and those dissolved in the serum rise to the top. These factors reduce Brownian motion and promote temporary agglomeration of particles. There is also a lowering of effective density for agglomerated particles and an increase in the effective size of the rubber particles and these two changes favor faster creaming.

Figure 1.4 Creamed latex with an upper creamed layer and a bottom serum layer

Both field latex and skim latex, which are byproducts obtained during the centrifugal concentration of fresh preserved field latex, can be subjected to creaming. Skim latex has an average particle size much lower than the average particle size of field latex. So, it is expected that creaming would be very slow. Skim latex can be creamed either by using a higher concentration (about 0.7%) of tamarind seed powder as a creaming agent or, alternatively, by employing a combination of alkali and fatty acid soap without a conventional creaming agent. In the latter method, creaming is facilitated by an increase in the density of the serum and an increase in the density difference between the serum and rubber network. Adsorption of fatty acid soaps onto the surface of rubber particles also plays a crucial role in the creaming process.

When the latex is concentrated without the application of centrifugal force, the rubber particles are subjected to three different forces: gravity (F_g), viscous drag (F_d) and buoyancy (F_b). During the centrifugal concentration of latex, the gravitational force is replaced by centrifugal force. For the mathematical derivation of the creaming process, we shall consider the forces F_d, F_g and F_b.

Assuming the rubber particle to be spherical:

$$F_g = 4/3\pi\, r^3\, \rho_r g \tag{1.1}$$

Here, ρ_r is the density of the rubber particle, r is the radius of the rubber particle, g is the acceleration due to gravity.

Viscous drag is proportional to viscosity η of the latex serum. If r is the radius of the rubber particle, v is the velocity of upward movement of the rubber particle through latex serum, the force due to viscous drag is given by the Stokes formula:

$$F_d = 6\pi\eta rv \tag{1.2}$$

The upward force exerted on the rubber particle by the fluid medium is called the buoyancy force (F_b).

(see Section 1.2.4.4.1)

$$F_b = \frac{4}{3}\pi r^3 \rho_s g \tag{1.3}$$

The drag force balances the difference between gravitational and buoyancy forces in the equilibrium condition. The equation governing concentration under gravitational force can be given as follows:

$$v = \frac{2g(\rho_s - \rho_r)r^2}{9\,\eta} \tag{1.4}$$

Here, v is the velocity of creaming (m/s), g is the acceleration due to gravity (m/s^2), ρ_s and ρ_r are the densities of serum and rubber particles, respectively (g/cm^3), r is the effective radius of the rubber particle (cm) and η is the viscosity of serum (centipoise).

Method of creaming

The primary creaming agents commonly used are ammonium and sodium alginates. Tamarind seed powder, a naturally obtained material, can also act as a creaming agent, as it contains some alginates. The steps involved are as follows.

1. Preserve latex with 1% ammonia and age preferably for 1–2 weeks

2. Remove magnesium ions present in latex by treating them with diammonium hydrogen phosphate (DAHP), which is added in a stoichiometric ratio for reaction with the magnesium content present in latex. (The quantity of DAHP required is 5.7 times the magnesium content). The quantity is 228 g of DAHP for one barrel of latex (200 L latex) if the magnesium content is 200 ppm (or 200 mg/L). Add DAHP as a 10% solution. In the reaction of DAHP with magnesium ions under the conditions prevailing in latex, magnesium ions are converted to magnesium ammonium phosphate and settle at the bottom.

   The Mg ions in latex interact with ammonia and phosphate anions, forming very sparingly soluble MgNH$_4$PO$_4$, which separates as a residue.

   $$Mg^{2+} + NH_3 + HPO_4^{2-} \rightarrow MgNH_4PO_4\downarrow$$

   Before concentration of the latex by either creaming or centrifuging, the Mg content of latex must be removed and the reaction period for magnesium removal is about 12 hours. Magnesium is removed as a sludge and is a source of phosphorous fertilizer.

3. Mixing of latex with creaming agents, both primary and secondary

Mix de-sludged latex with a primary creaming agent as a 3% solution in water. Since sodium and ammonium alginates are soluble in warm water, clear solutions of creaming agents are obtained by mixing with the necessary amount of warm water (between 50 and 60 °C), followed by cooling and straining. Add potassium/ammonium oleate soaps, the commonly employed secondary creaming agents. Commercial washing soaps are also used, particularly in a 10% solution. For 200 kg of latex, it is recommended to dissolve 100 g of soap in 1 L of water. Add soap solution to obtain a concentration of 0.05% soap, expressed in terms of latex. Mix the latex containing the creaming agents thoroughly for about one hour. Transfer the mixed latex to closed creaming tanks to prevent ammonia loss and allow the latex to remain undisturbed until the desired level of creaming is obtained. The minimum period is 48 hours. For the first 24–40 h, creaming is relatively rapid and then slows. Drain off the layer through the outlet valve at the bottom. Homogenize the latex concentrate and test the quality parameters based on standard specifications before using it commercially.

Centrifugation

When latex is subjected to high-speed rotation in a centrifuge bowl, the rubber particles experience centrifugal force in proportion to their mass. Heavier fractions experiences higher centrifugal force while lighter fraction experiences lower centrifugal force. Consequently, the fraction closer to the axis of rotation has comparatively more rubber and the fraction away from the axis of rotation will have a relatively low rubber content. Each layer is removed by means of annular spacing around the axis of rotation. Latex having a DRC of 30–35% can be concentrated to 60% or more by adjusting the length of the skim screw, which controls the retention time to which the latex is subjected to centrifugal force. As the length increases, the retention time of latex inside the bowl increases; consequently, rubber recovery in the concentrate will increase and the rubber content of skim latex will decrease. A longer retention time implies a lower latex flow rate and, consequently, a lower hourly output of the concentrate. Concentrated latex contains bigger rubber particles and the rubber molecules are highly branched, while skim latex contains smaller rubber particles and the rubber molecules are linear. Thus, it is possible to control the particle size distribution in the latex concentrate by controlling the skim screw length.

The design of the latex centrifuge is based on the conical plate centrifuge, also known as a disk stack separator or disk bowl. This type of centrifuge features a series of conical disks with suitable spacing between them. The space between two disks forms an individual centrifugation zone. The presence of many conical disks increases the surface settling area and the efficiency of centrifuging [38].

The conical plate latex centrifuge removes water from latex by applying a very high centrifugal force. In this process, the rubber particles, with a lower density than the serum, are compelled to move toward the center of the bowl. The most widely used type is Alpha Laval (LRB 510) from Sweden. The other machines are Westfalia from Germany and Westlake from China.

The machine features a rotating bowl that houses a set of concentric conical metallic separator disks, with an inclination ranging from 45–60°. It includes approximately 120 such disks, with stainless steel being the preferred material on account of its durability and resistance to corrosion. Each disk has a thickness close to 0.4 mm, which allows for efficient separation.

Latex fills the bowl before centrifugation through a feed tube and distributor. The separator disks with small holes placed one over the other in the shape of a cone and placed such that the position of the holes coincides in all disks and all the holes are at definite distances from the center of the axis of rotation. This configuration allows the latex to be distributed and broken into several thin conical shells within the bowl and facilitates collection of fractions with higher and lower rubber content separately through the holes and annular channels. The bowl containing latex rotates at a speed close to 7000 rpm during the centrifugation process. The centrifugal forces generated in the centrifuge are several thousand times larger than gravitational force, and this causes rapid separation of the latex into cream and skim. A photograph of a centrifuging unit is shown in Figure 1.5.

When the machine runs, the cream, a lighter fraction, moves inward toward the axis of rotation, while a heavier fraction with less dry rubber content, called skim latex, moves away from it. In the conical plates, there are holes and the cream fraction moves up through the holes, while the skim fraction has a channel near the periphery and moves toward the top through this channel. Sludge settles on the inner surfaces of the bowl and distributor, where it can be manually removed during cleaning.

Latex must be ammoniated and kept overnight before centrifuging; this allows the sludge to settle. The DRC of centrifuged latex can be regulated by adjusting the skim screw.

The efficiency of the centrifuging process (E) is defined as the proportion of total rubber recovered as concentrate.

$$E = W_c \times D_c / W_f \times D_f \tag{1.5}$$

Where

W_c = Weight of the cream

D_c = DRC of cream

W_f = Weight of field latex

D_f = DRC of field latex

(a)

(b)

Figure 1.5 (a) Centrifuging unit and (b) Centrifuging bowl with conical disks (Courtesy: PLPC, Rubber Board)

Usually, the efficiency attained in commercial units is 0.85–0.90, meaning that for every 100 kg of rubber processed, 85–90 kg will be obtained as concentrate. Factors which affect efficiency are feed rate, angular velocity of the machine, length of regulating screws and DRC of field latex. Efficiency increases with reduction in the feed rate of field latex or increase in the speed of the centrifuge. A shorter screw increases the DRC of latex concentrate, but reduces the efficiency of the centrifuging process. The DRC of latex and the size of rubber particles can be controlled by adjusting the skim screw. A longer skim screw gives centrifuged latex a higher proportion of larger-sized particles and, consequently, skim latex of a higher DRC.

The rubber particles in the cream fraction of centrifuged latex are mostly large rubber particles (LRPs). The rubber molecules in large rubber particles are highly branched. Centrifuged latex has a bimodal particle size distribution. The skim screw length of the centrifuge can be varied to yield centrifuged latex skim latex with different particle size ranges, as illustrated in Figure 1.6. The centrifuged latex labeled 1 has a higher proportion of larger particles than those labeled 2 and 3. Molecules in bigger particles are highly branched and those in smaller particles are linear. Better green strength and gum strength are observed in highly branched rubber molecules than in linear ones [16, 39].

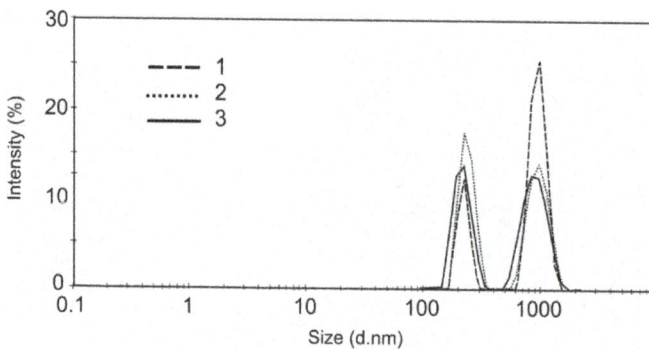

Figure 1.6 Effect of skim screw length on the particle size of centrifuged latex

During centrifugal concentration, a byproduct called skim latex is produced in almost the same quantity as latex concentrate. The centrifuged latex contains mainly LRPs, while the skim latex has SRPs. The rubber molecules in SRPs are linear compared with LRPs, where molecules are highly branched on account of phospholipids at the chain end of the rubber molecules. Linear rubber molecules are believed to exhibit chain entanglements and, after vulcanization, protein-removed skim rubber can possess better mechanical properties. Due to the absence of branching, skim rubber does not show the phenomenon of storage hardening, which is normally observed in conventional sheet rubber.

Skim latex contains 2–5% rubber and is processed to collect the rubber known as skim rubber. Skim latex contains small rubber particles, more adsorbed proteins and less acid coagulation sensitivity. The recovered rubber contains many proteins and fatty acids and has a higher modulus and poor dynamic mechanical properties. The quality of skim rubber can be improved by suitable protein removal techniques to obtain rubber with better mechanical and dynamic properties. The proteins of skim rubber can be removed by enzymatic [40] or alkaline hydrolysis of proteins [41], followed by creaming.

The effluent is subjected to processes such as sedimentation, aerobic or anaerobic oxidation and other suitable treatments to yield the required biological oxygen demand (BOD), chemical oxygen demand (COD) and other parameters per international standards.

The BIS (Bureau of Indian Standards) specifications for preserved latex concentrates are contained in IS 5430–2017 for centrifuged latex (type high ammonia, low ammonia and medium ammonia) and in IS 13101–1991 for creamed, ammonia-preserved latex.

Double centrifuged latex contains low levels of non-rubber ingredients, mainly proteins. A very low protein content is required for eliminating problems related to protein allergy and improve rubber insulation behavior. For this reason, double centrifuged latex is useful in producing products such as condoms, surgical gloves, electrician's gloves and catheters. In the production of double centrifuged latex, the single centrifuged latex is diluted to about 35–40% DRC and recentrifuged to 60% DRC. Specifications for double centrifuged latex are prescribed in IS 11001:1984.

Prediction of quality parameters of centrifuged latex

The critical quality parameters of latex concentrate are DRC, total solids content (TSC), alkalinity, sludge content, mechanical stability time, coagulum content, mechanical stability test (MST) and volatile fatty acid (VFA) number. The MST of latex is generally low when freshly prepared and increases in storage. With reasonable accuracy, an accelerated aging test can predict important quality parameters, such as MST and viscosity. The test is done by subjecting latex to accelerated aging under 1% ammonia at 50 °C for 96 hours and measuring the quality parameters. The quality parameter after accelerated ageing will be in reasonable agreement with the quality parameters attained after one month storage of centrifuged latex at room temperature (Table 1.2 and Table 1.3).

The MST of latex increases in storage because of the fatty acid soaps generated in the latex. Under accelerated aging conditions, this reaction is expected to happen.

Evaporation: In a typical process, latex is circulated through a tubular heat exchanger and then passed into a chamber where evaporation occurs under reduced pressure.

Electrodecantation: Under an applied potential difference between two electrodes in a cell, the negatively charged rubber particles move toward the anode and the latex acquires a higher DRC (seldom used).

Table 1.2 The Quality Parameters of Latex Concentrate (High Ammonia) after Accelerated Aging and Storage for One Month

Sample No.	Time Frame for Testing Quality Parameters					
	One Day after Production		After Subjection to Accelerated Aging		After Storage for One Month at Room Temperature	
	MST (s)	Viscosity (cps)	MST (s)	Viscosity (cps)	MST (s)	Viscosity (cps)
1	90	144	380	120	390	122
2	105	136	840	96	810	100
3	201	160	650	112	622	114
4	121	168	475	116	463	114
5	62	142	590	122	600	120
6	120	88	630	84	650	86
7	110	108	350	92	370	105

Table 1.3 The Quality Parameters of Latex Concentrate (High Ammonia) Obtained after Accelerated Aging and Storage for One Month, Containing Various Concentrations of Fatty Acid Soap

Concentration of Potassium Laurate Solution, % (m/m)	Time for Testing Quality Parameters					
	One Day after Production		After Subjection to Accelerated Aging		After Storage for One Month at Room Temperature	
	MST (s)	Viscosity (cps)	MST (s)	Viscosity (cps)	MST (s)	Viscosity (cps)
0.0	187	134	1103	114	1200	108
0.005	202	132	1300	110	1310	105
0.010	248	129	1495	108	1440	105
0.015	295	128	1690	106	1800	103
0.020	379	122	1980	101	2040	98
0.025	428	116	2306	100	2340	93
0.030	563	115	2500	96	2610	90

1.2.2 Sheet Rubber: Ribbed Smoked Sheet and Air-Dried Sheets

Ribbed smoked sheet rubber, pale latex crepe rubber and specific grades of techni-cally specified rubber (TSR) are obtained from latex. Marketable dry rubber forms include estate brown crepe rubber and TSR produced from field coagulum. The field coagulum contains all the solids content of latex. By contrast, in sheet rubber, much of the non-rubber solids are removed during the controlled dilution and acid coagu-lation conditions. Some non-rubber ingredients are retained on dry rubber after co-agulation and the main non-rubber ingredients are soluble in acetone; these are mainly lipids and sterol esters. A small quantity of carbohydrates is also retained on coagulated rubber. Inorganic substances are quantified as ash content. Non-rubber ingredients, such as lipids and proteins, affect vulcanization and aging characteris-tics.

They are produced by drying the shaped (grooved) rubber sheets obtained from suitably diluted latex, coagulating them in suitable containers and sheeting them through smooth and grooved rollers. The international market has two types of sheet rubber: ribbed smoked sheet (RSS) and air-dried sheet (ADS) (Figure 1.7). Air-dried sheets have a good yellow color, but are susceptible to being attacked by fun-gus. Of these two types, RSS is the more popular and is available for volume con-sumption.

Figure 1.7 Air-dried rubber sheets

Rubber sheets are produced from fresh natural rubber latex by a process of slow la-tex coagulation after addition of suitable coagulants, such as formic acid or acetic acid, to latex.

The various steps involved in the production of rubber sheets are dilution of latex to a particular dry rubber content, addition of chemicals and coagulants that lead to slow coagulation of latex, conversion of wet coagulated mass to rubber sheets and drying of same in smoke to yield ribbed smoked sheets. The details are presented below.

1.2.2.1 Sieving and Dilution of Latex

For processing, latex collected in clean containers is brought to the factory. Cleanliness is essential for preventing degradation of the latex quality. Coconut shells (popularly used) or plastic cups of high-density polyethene collect the latex from the tree. Buckets and storage tanks made of aluminum or galvanized iron coated with an inert paint are used to maintain the quality of latex.

Normally, field latex coagulates after four to six hours of tapping. It has a tendency to pre-coagulate, a process observed in certain conditions, such as the influence of wintering, clonal variations and contamination with rainwater while the latex is being collected from the tree. A few drops of anti-coagulants can be added to the collection cup or after collection from the tree. Chemicals added to latex to prevent pre-coagulation are sodium sulfite, ammonia or formalin.

Sodium sulfite is used in a concentration of 0.05% v/v. An amount of 5 g of the chemical in 100 mL water is sufficient for 10 L field latex. Ammonia is used as liquified gas in cylinders or as a 20% solution in water and used at a concentration of 0.01% v/v. A volume of 5 mL of the solution in 100 mL water is sufficient for 10 L field latex. Formalin is available as an approximately 40% solution of formaldehyde in water and is used in a concentration of 0.02% v/v. A volume of 5 mL of formalin in 100 mL water is sufficient for 10 L field latex.

Fresh latex is passed through 40-mesh sieves to remove dirt and is then diluted in bulking tanks to a DRC of about 12.5–14%, so as to get about 500 g of dry rubber sheet for every 4 liters of diluted latex. If all sheets have similar dried weights, drying wet sheets in a smokehouse would be easier. The diluted latex can stand in the bulking tank for 15–20 minutes to facilitate sedimentation of heavy dirt particles.

The quantity of water required for diluting the latex to 12.5% DRC is calculated from Equation 1.6:

$$V = \frac{D \times V}{12.5} - v \qquad\qquad (1.6)$$

Where

V = Volume of water in liters (L) to be added to field latex

D = DRC of latex and

v = Volume of latex in liters

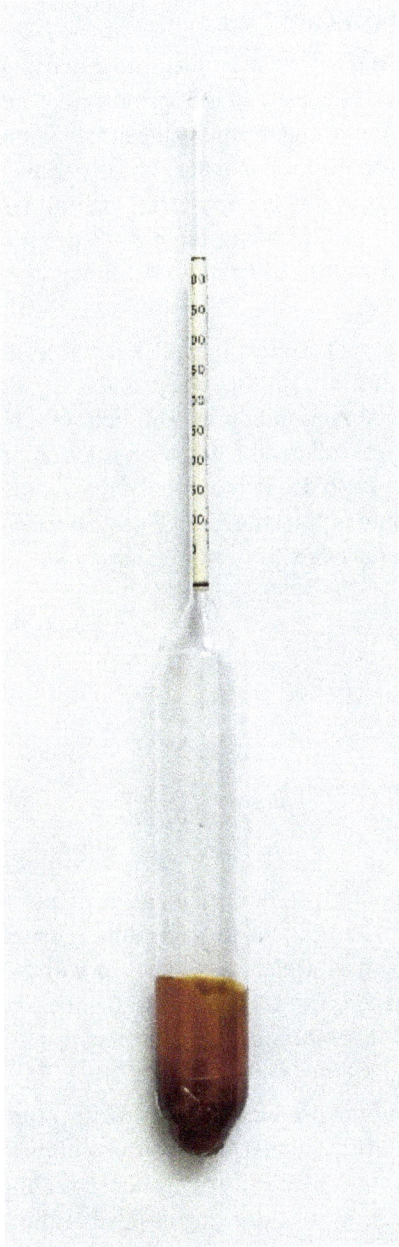

Figure 1.8 Metrolac

1.2.2.2 Estimation of Dry Rubber Content (DRC) of Latex

Metrolac method: The dry rubber content of field latex is determined roughly using a hydrometer known as the 'Metrolac' (Figure 1.9). This method is approximate; DRC variation can be 3–10%. However, it is helpful in the primary processing of latex into ribbed sheets. Latex is diluted with water in a proportion of one part latex with two parts water (1:2 proportion). The diluted and mixed latex is transferred to a tall jar and, once the bubbles have subsided or been removed carefully, the metrolac is immersed in it. The reading on metrolac is read and multiplied by three if dilution is in the ratio 1:2 or by two if dilution is in the ratio 1:1.

Microwave oven drying method: This is an alternate method for quick determination of DRC by drying the coagulum in a microwave oven. For this, weigh about 10 g of natural rubber latex and add about 0.8–1 mL of ammonium laurate solution. Admix 0.8% sulfuric acid until the latex coagulates to a consolidated mass after addition of acid. Wash the coagulum, press it into a thin sheet and dry in a microwave oven for 10–12 minutes (set at 360 watts). After about 5 minutes, turn the sample upside down. After drying, take the weight of the dried rubber. Calculate DRC from Equation 1.7.

$$DRC\ (\%) = \frac{W_1 \times 100}{W_2} - C \tag{1.7}$$

Where

$C = 1$ for ammonium laurate

W_1 = Weight of the coagulum after drying in the microwave oven

W_2 = Weight of the latex

If the coagulum is further dried at 100 °C for about half an hour in an air oven until constant weight of the coagulum is obtained, it can be assumed the coagulum is dried correctly. The DRC calculated using this weight as dried weight would be close to the DRC calculated using the standard laboratory method. The error of the DRC obtained in this way would be within 0.1% of the standard laboratory method and can be considered as a substitute for the standard laboratory method (Section 1.2.4.4.1).

Dilution helps achieve quality consistency for the final product. It also yields a softer coagulum, which can be easily sheeted. Dilution also facilitates faster settling of dirt, thereby improving the clarity of the sheets, lowering the viscosity of the latex, distributing the chemicals added to the latex more uniformly and allowing trapped air and fermenting gases to escape from the latex more easily.

Latex from some clones darkens the final RSS sheet, due to enzymatic discoloration. If this discoloration is experienced, a sufficient quantity of sodium metabisulfite or sodium bisulfite should be added as a 3% w/w solution (1.2 g/kg DRC).

The commonly used coagulants are acetic acid and formic acid. Suitably diluted sulfuric acid may also be used. If so, the process of converting latex to sheet should be completed on the same day to prevent discoloration of the sheets. The acid requirement is given in Table 1.4. The use of salts such as calcium chloride and sodium chloride with formic acid can reduce the quantity of formic acid used for latex coagulation.

1.2.2.3 Latex Coagulation and Production of Sheets

The addition of acids in diluted form promotes uniform distribution within the latex, facilitating complete coagulation and yielding a soft coagulum suitable for producing sheets free from air bubbles and tackiness.

Table 1.4 Acid Requirement for Coagulation of 4 Liters of Diluted Latex Containing ½ kg Rubber

Time for Coagulation and Maturation	Acetic Acid	Formic Acid	Sulfuric Acid*
Next day	3 mL diluted to 150 mL with water	1.5 mL diluted to 150 mL with water	250 mL of 0.5% acid
Same day	4 mL diluted to 200 mL with water	2 mL diluted to 200 mL with water	300 mL of 0.5% acid

* Highly corrosive care should be taken in handling and dilution. Thorough washing of the coagulum during and after sheeting is essential to remove acid residues, which, if retained, can affect rubber processing.

As it is highly corrosive, care should be taken in handling and dilution. Thorough washing of the coagulum during and after sheeting is essential for removing acid residues, which, if retained, can affect rubber processing.

Four liters of diluted latex is transferred into aluminum pans to ensure the sheets are uniform. After adding the acid solution, the latex should be mixed thoroughly using an aluminum sheet or a wooden plate. The froth may contain dirt and should be removed. Removal of froth can prevent the formation of pitting of the surface of the dried sheets. This also prevents air entrapment inside the rubber sheets. The stacking of aluminum pans containing latex in large-scale production of rubber sheets is shown in Figure 1.9.

Figure 1.9 Arrangement of latex coagulation pans in a processing center

Figure 1.10 Dripping of sheets under shade

Figure 1.11 Typical furnace of a smokehouse

1.2.2.4 Defects in Rubber and Grading of Sheets

The sheets are graded by visual examination, normally accomplished by holding sheets against the light whereupon the most obvious defect becomes apparent. RSS1X, RSS1, RSS2, RSS3, RSS4 and RSS5 are the standard grades. The sheets should have an elastic nature, should not be opaque and should be of sufficient strength. Weak sheets and those attacked by fungus are graded as off-grade, which is shown in Figure 1.12.

Defects such as pinhead bubbles, rust, discoloration and mold growth are of microbial origin:

- Pinhead bubbles are small globules of air or gas. The surface inside is dry and not sticky. This may be caused by bacterial action arising from a longer coagulation time resulting from insufficient acid

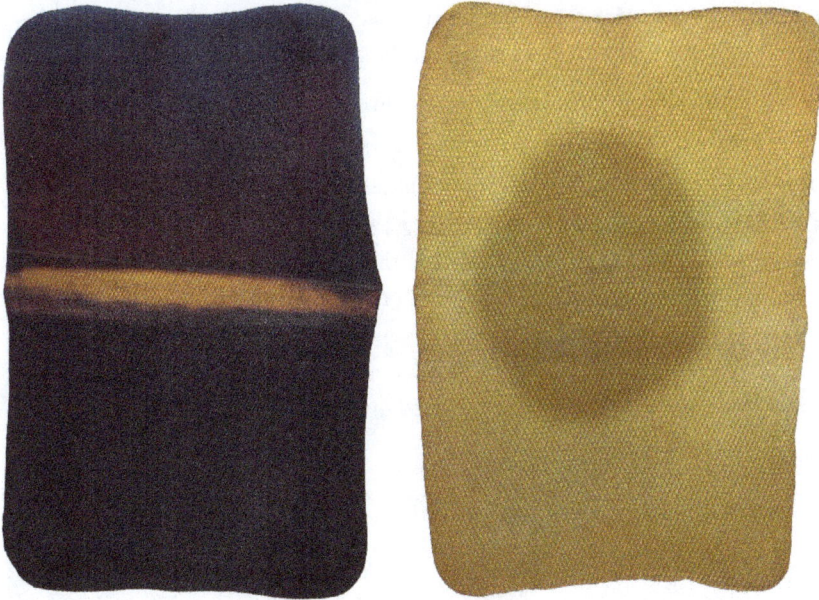

Figure 1.12 Typical images of off-grade sheets

- Rust is a readily visible non-rubber brownish deposit present on the surface of smoked sheets, without the sheets' having been stretched or scratched. It may be caused by prolonged dripping of wet sheet that allows microbial action mainly by yeast and bacteria. This defect should not be confused with mold growth; the latter can be removed largely by brushing, but rust remains firmly attached to the sheet

- Discoloration is a color change in sheets, due to enzymatic action on the non-rubber constituents present in latex. Certain phenolic compounds in latex in the presence of oxygen and an enzyme such as polyphenol oxidase (PPO) produce diphenols; these are oxidized to quinones, which react with amino acids or proteins to form brown-colored products, causing discoloration in rubber sheets. This discoloration can be prevented by adding sodium bisulfite, a more potent reducing agent than PPO. Oxygen can react with sodium sulfite and thus is unavailable to form colored compounds

- Staining can also happen as a result of rubber's biochemical degradation, which may occur when sheets are packed before drying

- Mold growth is mainly due to improper drying of sheets

- Case hardening is a white streak in the cut section of the sheet for any reason, such as rapid drying of sheets in early stages, insufficient dilution of latex and excessive thickness of sheets. Under these circumstances, the outer portion dries and the entrapped moisture does not even escape if the temperature increases

- Tar drops are black drops resembling tar and are due to the condensation of water vapor carrying carbon deposits

- Stickiness, due to excess acid or drying at higher temperatures

Sheets are graded and marketed as 50 kg or 111.1 kg bales for the international market and packed in polyethene sheets of suitable thickness, density and melting point. The Bureau of Indian Standards (BIS) specifications for various marketable forms are given in Table 1.5.

Table 1.5 Specification for Natural Rubber Based on BIS

Indian Standards Natural for Rubber (ISNR)	IS 4588–1986
Ribbed smoked sheet (RSS)	IS15361–2003
Centrifuged latex	IS 5430–2017
Double-centrifuged latex	IS 11001–1984
Creamed latex	IS 13101–1991

1.2.3 Block Rubber or Technically Specified Rubber (TSR)

Crepe rubber and sheets are graded by visual examination. In the consumer market, other examination methods for grading may be more attractive and rubber packaged in solid block form can be more appealing. Technically specified rubber (TSR) was introduced to strengthen the competitiveness of natural rubber against synthetic alternatives in terms of presentation and grading. The field coagulum is immersed in tanks of water for 1 or 2 days, the exact duration depending on the type of coagulum.

As stated below, at least ten passes through various machines are required for size reduction. Coagulum cutters, slitting knives, slab cutters (Figure 1.13), shredders (Figure 1.14), pre-breakers (Figure 1.15), creper mills and hammer mills are the types of machinery used for size reduction. A slab cutter is primarily used to cut and break the field cup lumps and rubber blocks. This machine consists mainly of a heavy-duty rotor with fixed knives working in tandem with fixed knives on the machine's housing. A pre-breaker consists of two scrolls with cutting edges. When the scrolls rotate, the rubber lump (obtained from a slab cutter) is pulled apart or breaks apart and is forced through a die plate. In other words, the machine works like an extruder. The coagulum obtained from a creping machine in the form of a fine blanket can be converted into a wet crumb with a shredder. This is a high-speed cutting machine whose main purpose is final size reduction of crepe blankets obtained from latex coagulum.

A hammer mill has hardened steel hammers that rotate inside a chamber at a high speed. The rubber lump inside the chamber is thus cut into crumbs by the radially swinging hammers.

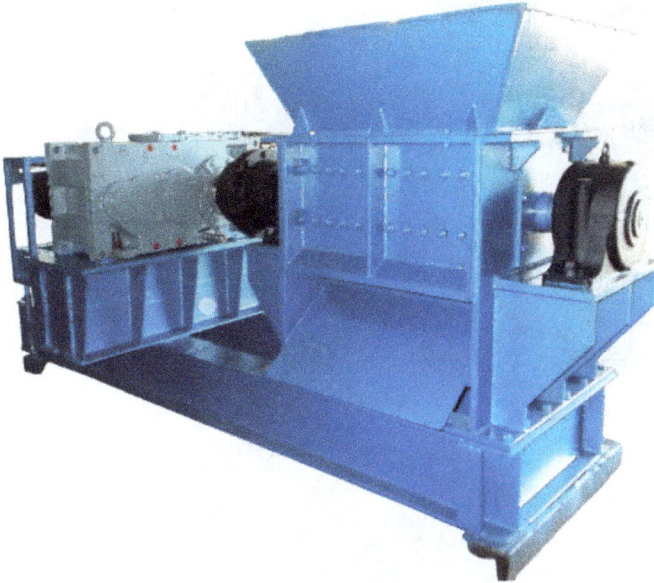

Figure 1.13 Slab cutter

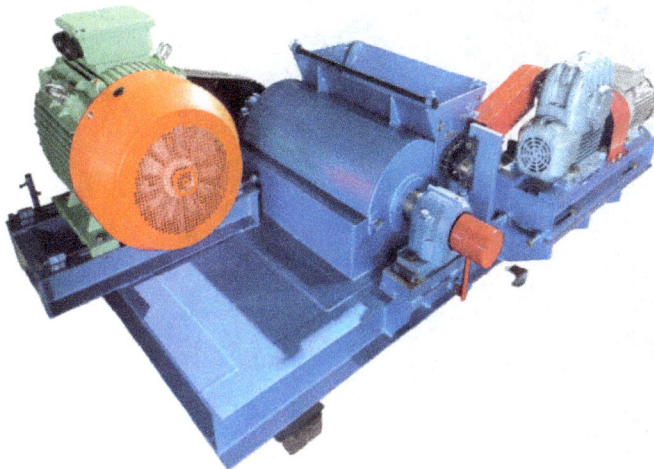

Figure 1.14 Shredder (Courtesy: Hevea Engineering Works, Kerala, India)

Size reduction with a pre-breaker is shown in Figure 1.15.

(a)

(b)

Figure 1.15 Size reduction with a pre-breaker: (a) side view and (b) sample subjected to size reduction (Courtesy: Pilot Crumb Rubber Factory, Rubber Board, India)

Figure 1.16b shows size reduction on a creper mill. The size of the field coagulum after immersion in water for a specific time and washing or of latex coagulum after coagulation, possibly by pit coagulation, is reduced by passing through several items of equipment.

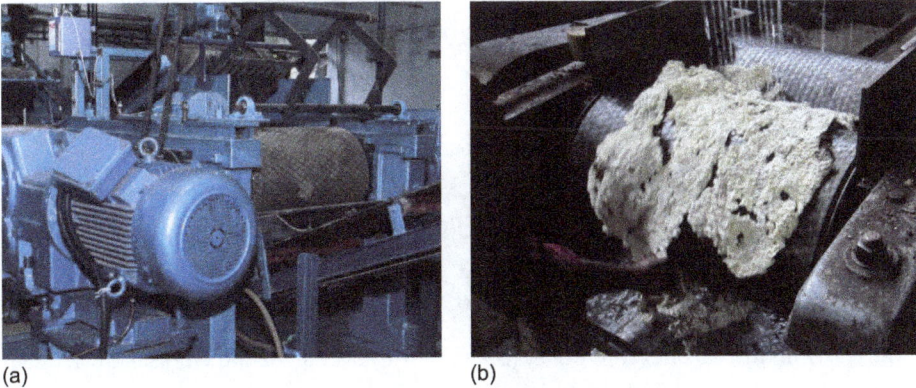

(a)   (b)

Figure 1.16 (a) Creper mill and (b) size reduction of rubber using creper mill (Courtesy: Hevea Engineering Works, Kerala, India)

A creper mill is a two-roll mill with an individual motor and gearbox. The rolls have diamond-shaped cuttings and medium grooves. Macerators are similar machines that are used to turn field coagulum into a thick blanket. In contrast to creper mills, macerators have deeper grooves for the diamond-shaped cuttings.

The final size reduction is achieved with a shredder or hammer mill to afford rubber crumb. The crumb is washed well and dried at 70–100 or 115 °C in automatic dryers (Figure 1.17). The drying time is close to 4–6 hours. The dried crumb is then pressed into 25 kg bales by a hydraulic press (Figure 1.18).

The dried crumb is pressed into 25 kg bales by a hydraulic press. Samples are cut in a few representative bales and the quality parameters tested for compliance with national specifications. Technically specified rubber (TSR) or block rubber is graded according to BIS specifications, IS 4588: 1986 (Reaffirmed Year: 2018). The test parameters and corresponding requirements are presented in Table 1.6 [42]. There are six grades for Indian Standard Natural Rubber (ISNR), namely ISNR3CV, ISNR3L, ISNR5, ISNR10, ISNR20 and ISNR50. Rubber is packed in polyethylene sheets and marked with grade, net mass, producer, month and year of production. Sheet rubber and TSR are equally suitable for producing many rubber products, including tires.

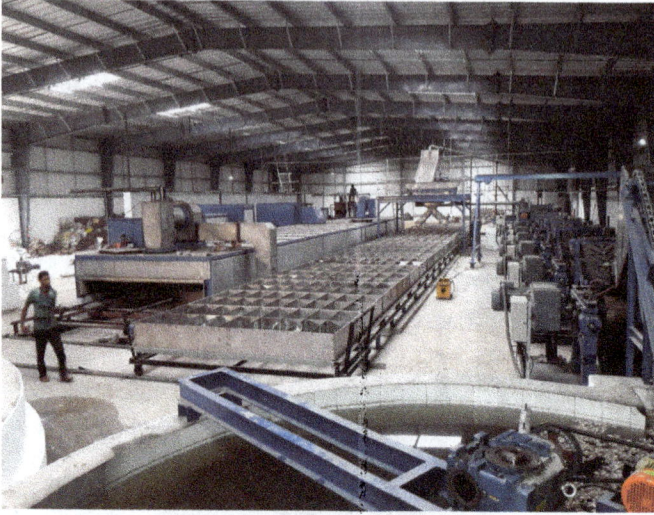

Figure 1.17 Industrial drier with trolley (Courtesy: Hevea Engineering Works, Kerala, India)

Figure 1.18 Hydraulic press (Courtesy: Hevea Engineering Works, Kerala, India)

Table 1.6 Indian Standard Specification for Raw Natural Rubber

Parameter	Requirement for					
	ISNR3CV	ISNR3L	ISNR5	ISNR10	ISNR20	ISNR50
Dirt content percent by mass (max.)	0.03	0.03	0.05	0.10	0.20	0.50
Volatile matter percent by mass (max.)	0.08	0.80	0.80	0.80	0.80	0.80
Ash, percent by mass (max.)	0.50	0.50	0.60	0.75	1.0	1.50
Nitrogen, percent by mass (max.)	0.60	0.60	0.60	0.60	0.60	0.60
Initial plasticity (min.)	As agreed between the purchaser and supplier	30	30	30	30	30
Plasticity retention index (min.)	60	60	60	50	40	30
Color Lovibond	-	6.0	-	-	-	-
Mooney viscosity ML 1+4 at 100 °C	60 ± 5	-	-	-	-	-
Accelerated storage hardening	8	-				

The sampling and homogenization procedure for testing raw rubber is presented in Section 1.2.4.4.2.

1.2.4 Crepe Rubber

Crepe rubbers are processed from latex coagula, field coagula or cuttings of RSS. After preliminary treatments, coagulum from fresh latex or any form of field coagulum is passed through a set of creping machines to become lace-like rubber, so-called on account of the marks produced by the machine, which is then dried to form crepe rubber. Crepe rubbers are classified as latex crepe or field coagulum crepe. Pale latex crepe (PLC) is obtained from field latex, while estate brown crepe is obtained from field coagulum. Different grades of PLC are used for products such as pharmaceutical products, light and brightly colored goods, adhesives, tubing, etc. Sole crepe, primarily developed for shoe soles, is now preferred for light-colored products and adhesives.

1.2.4.1 Pale Latex Crepe

The latex used to prepare PLC must be free from yellow pigments as far as possible. The coagulum must not undergo discoloration. To prevent pre-coagulation, sodium sulfite (0.05%, calculated on latex) or ammonia (0.01%, calculated on latex) may be added. As formalin can activate enzymatic discoloration, it cannot be used as an anti-coagulant. The latex is sieved through both a 40- and a 60-mesh sieve to remove impurities, the DRC is determined and the latex is diluted to a DRC of 25%. Chemicals are added to prevent enzymatic discoloration. Sodium bisulfite or sodium metabisulfite is used for this purpose. The chemicals are used at a dosage of 0.5%, calculated on DRC of latex, and added as a 5% solution. The latex is stirred well after addition of the chemicals and is kept undisturbed to denser impurities to settle.

The coloring agents can be removed by partial coagulation, addition of bleaching agents or a combination of both.

Preferential coagulation

In preferential coagulation, a suitable quantity of coagulant is added, stirred for 20–30 minutes and left for some time to form a coagulum containing the coloring materials. The coagulants may be acetic acid or oxalic acid. Acetic acid is added at a rate of 0.625–1.1 mL/kg dry rubber in the form of a 1% solution or, alternatively, oxalic acid is used at 0.625 g/kg dry rubber as 2% solution. The acid-treated latex is allowed to stand for 1–2 hours and is then sieved through 60 mesh to remove the fraction of latex containing yellow pigment, which roughly comes to 10–15% of latex's total dry rubber content. The remaining latex is coagulated with formic acid (3.3–4.4 mL/kg dry rubber as a 1% solution) or oxalic acid (5.6–7.5 g/kg dry rubber as a 2% solution). The use of oxalic acid offers better color and better color retention.

The process has the advantage of being cheap process and the disadvantage of being time-consuming, with only a part of the crop converted into the top-quality crepe.

Bleaching

This is the preferential coagulation of latex-coloring materials and a small portion of rubber from the residual latex. This technique uses a chemical that preferentially reacts with the coloring materials present in latex to remove the color. The commonly available bleaching agent used earlier was RPA No. 3 (xylyl mercaptan), now banned for health reasons. Other bleaching agents used were Cureobleach (tolyl mercaptan), which has a foul odor and is harmful to health, and Nexobleech, which is a sodium salt of toluene parathiol. These chemicals can be added in a concentration of 1.9–2.5 L per 100 kg dry rubber in the form of a 5% solution. A better color for the crepe is obtained if the bleaching agent is added after fractional coagulation. The coagulum obtained is milled in a crepe-making machine and dried at room temperature.

The advantages of using a bleaching agent are faster processing and complete conversion of latex into a top-quality crepe. The disadvantages are the high cost of the bleaching agent and the harmful effect on the rubber if the agent is used in excess quantity.

1.2.4.2 Field Coagulum Crepe

Along with latex, the main crop, field coagula such as cup lumps and tree lace are also harvested. If the field coagulum is processed while fresh, the resulting crepe rubber, also known as estate brown crepe (EBC), will be of superior quality. EBC can be used in mechanical and extruded goods and products, such as mats. Crepe from field coagulum is classified into five types, based on the type of coagulum and its blending with some other form of rubber: (1) estate brown crepe, (2) thin brown crepe, (3) thick blanket crepe, (4) flat bark crepe, and (5) pure smoked blanket crepe. For the production of crepe rubber, the coagulum is passed through different machines, such as a macerator and crepe roller. The number of passes through these machines depends on the type of coagulum.

1.2.4.3 Reclaimed Rubber

Natural virgin rubber is biodegradable, but not considered a biodegradable material after vulcanization. The recycling of used rubber products and waste rubber from rubber factories is necessary from an environmental perspective.

One method is to grind the rubber to a fine powder and reuse it by mixing it with raw rubber. Tire buffings and powdered rubber obtained from waste microcellular rubber products are two materials that serve as lightweight fillers in conventional rubber compounds. Latex-based waste rubber in raw form or products such as gloves, mainly rejected by industry, can be masticated and mixed with raw rubber in small proportions of 5–10%. The quality of this raw material is high, as its molecular weight remains high because the processing of latex products does not involve any mastication.

Natural rubber exhibits limited resistance to thermal and oxidative aging. Under accelerated aging conditions, the polymer undergoes degradation through scission of the main chain or the cleavage of vulcanization-induced cross-links. The reclamation of rubber is achieved by inducing controlled chain scission using thermo-mechanical or mechano-chemical processes, which selectively break molecular bonds to facilitate reprocessing. Scrap rubber products, mainly old tires, tubes, seals, gaskets and similar products, either alone or in the presence of strong acids, such as sulfuric acid, or of strong alkalis, such as sodium hydroxide, are heated to a high temperature by steam to yield devulcanized rubber. The heater process patented in 1885 by Hiram L. Hall, the acid process patented by N. Chapmann Mitchell in 1881 and the alkali digester process developed by Arthur H. Marks in 1899 were based on these techniques [43].

Mechanical work done on an extruder and the application of reclaiming chemicals made reclamation easier. The popular reclaiming agents were diphenyl disulfide, diamyl disulfide, butyl mercaptan and thiophenols.

The main benefits of using reclaimed rubber are cost reduction in a rubber product, enhanced processability characteristics, such as reduced die swell, reduced shrinkage, faster curing (depending on the type of reclaim), and enhanced mechanical properties, such as improved fatigue resistance.

Reclaim rubber is used in bicycle tires, passenger vehicle tires, off-the-road tires (OTR), conveyor belts and general mechanical products.

1.2.4.4 Retreading of Tires and Reuse of Waste Tires

The growing reuse of tires has led to a corresponding rise in the use of vehicle tires, which is accompanied by an increase in the conversion of used tires into waste tires. When burned, the rubber additives and components in tires release toxic chemicals, contributing to air and water pollution, and are therefore considered hazardous. Tires are reused through methods such as retreading, mechanical or cryogenic pulverization [44, 45], and repurposing the resulting material. Pulverized rubber is used in road surfacing blends with bitumen, mixed with soil to obtain specific advantages or employed as a filler in rubber compounds.

The different marketable forms of natural rubber find application in general rubber goods. However, specific grades can be selected on the basis of cost and property requirements. RSS 1X and ISNR3L are premium grades of natural rubber and can be used in aero tires. PLC 1X and ISNR3 CV are used for light-colored products, such as injection bottle caps, latex bulbs and medical tubes. RSS1, PLC and ISNR5 can be used in natural rubber-based tubes, food conveyor belts and engineering products, such as shock absorbers and elastomer bearings. RSS2 and ISNR10 can be used in premium-grade footwear, hoses and radial tires. General tires, tubes, hoses and footwear can be made from RSS3, RSS4, ISNR10 or ISNR20. RSS 5, ISNR50 and EBC can be used for general-purpose hoses, cycle tires and general mechanical goods.

1.2.4.4.1 Appendix A

1. Mathematical relations involved in the creaming of latex

The upward force exerted by water pressure on the object is the buoyancy force. An object floats when the buoyancy force equals its weight: $F_b = mg$, where m is the mass of rubber particle and g is acceleration due to gravity. The buoyancy force acts against gravity and pushes immersed objects upward. Its magnitude depends on the fluid's density – the denser the fluid, the greater the buoyancy force. For a given density difference, the buoyancy force on a particle increases with the third power of the particle diameter. In contrast, the viscous drag on the particle increases only with the first power of the diameter. The work required to move a particle a given distance down through the aqueous phase increases with the third power of the diameter.

$$P = F/A \tag{1.8}$$

$$F = P \times A \tag{1.9}$$

P, the pressure exerted by a water column of height h, is given by

$$P = \rho g h \tag{1.10}$$

Here, ρ is the density of the fluid, h is the height of the fluid column, g is the acceleration due to gravity.

$$F_b = P_{bot} A - P_{top} A \tag{1.11}$$

$$F_b = (\rho g h_{bot}) A - (\rho g h_{top}) A \tag{1.12}$$

$$F_b = \rho g A (h_{bot} - h_{top}) \tag{1.13}$$

$(h_{bot} - h_{top})$ can be considered as the height if the object is cubic.

Hence,

$$F_b = \rho g A h \tag{1.14}$$

Where

Ah = volume of the object

$$F_b = \rho_r g V \tag{1.15}$$

Where

V = volume of fluid displaced

However, here, the rubber particle is in the aqueous medium, so V can be taken as the volume of the rubber particle.

For spherical particles,

$$\text{Volume} = \frac{4}{3}\pi r^3 \tag{1.16}$$

$$F_b = \frac{4}{3}\pi r^3 \rho_s \, g \tag{1.17}$$

Where

ρ_r = density of the rubber particle

ρ_s = density of the serum

r = radius of the rubber particle

g = acceleration due to gravity

$$F_g = \frac{4}{3}\pi r^3 \rho_r \, g \tag{1.18}$$

Due to the viscosity of aqueous serum, rubber particles can experience a viscous drag, because viscosity is a property of fluid that opposes the shearing of the fluid. For a solid sphere of radius r moving at speed v in a fluid of viscosity η, the viscous drag is given by the Stokes equation:

$$F_d = 6\pi\eta r v \tag{1.19}$$

At equilibrium, the drag force balances the difference between gravitational force and force due to buoyancy.

$$6\pi\eta r v = \frac{4}{3}\pi r^3 g - \frac{4}{3}\pi r^3 \rho_r g \tag{1.20}$$

$$v = \rho_s 2g(\rho_s - \rho_r)r^2/9\eta \tag{1.21}$$

2. Standard laboratory method for the determination of dry rubber content (DRC)

About 10–15 g of latex is accurately weighed into a beaker. The latex is diluted with about an equal volume of water and 15–20 mL of 1% formic acid. The latex is stirred well and left aside for 15–20 minutes until latex coagulates to a smooth coagulum, leaving clear serum. The coagulum is washed well, pressed into a thin sheet and dried at 70 °C in an air oven until constant values are obtained.

$$DRC = (w_r/w_l) \times 100 \tag{1.22}$$

Where

w_r = Weight of dry rubber

w_l = Weight of latex

1.2.4.4.2 Appendix B
Sampling procedure

Place the rubber bale on a clean surface, with the shortest edges vertical. Cut a sub-piece of a triangular section (about 50 x 50 x 40 cm) down the entire length with a minimum weight of 180 g. Cut another similar piece from the diagonally opposite corner. These two sub-pieces are used to test the quality parameters of ISNR and it is assumed that this piece represents the ISNR bale.

Homogenization

The sample is processed by passing six times through the gap between the rolls of a laboratory mill measuring 150 x 300 mm (6 x 12 inches), with the rolls rotating at different speeds in a ratio of 1:1.4. To maintain the temperature, the rolls are cooled with running water at room temperature. The gap is set at 1.65 mm (0.065 inches). After each pass, the rubber is rolled into a cylinder and introduced end-wise for the next pass. The rubber is not rolled after the sixth pass.

The homogenized rubber is subjected to various tests as per the parameters mentioned in IS 4588.

Notations Used

F_g	Force due to gravity
F_b	Force due to buoyancy
F_d	Force due to viscous drag
r	Radius of rubber particle
ρ_r	Density of rubber particle
ρ_s	Density of latex serum
V	Speed of movement of rubber particles in latex
η	Fluid (latex serum) viscosity
DRC	Dry rubber content

References

[1] K. P. Jones, P. W. Allen, 'Historical Development of the World Rubber Industry', in *Developments in Crop Science*, vol. 23, Elsevier, 1992, pp. 1–25. doi: 10.1016/B978-0-444-88329-2.50007-6.

[2] P. J. George, Vijayakumar, K. U. Thomas, Rajagopal, C. Kuruvilla Jacob, Eds., 'Chapter 12', in *Natural rubber: agromanagement and crop processing*, Kottayam, India: Rubber Research Institute of India, 2000.

[3] H. Y. Yeang, O. Hashim, 'Destabilization of Lutoids in Hevea brasiliensis Latex During Early and Late Flow', vol. 11, 1996.

[4] X. Gidrol, H. Chrestin, H. L. Tan, A. Kush, 'Hevein, a lectin-like protein from Hevea brasiliensis (rubber tree) is involved in the coagulation of latex', *J. Biol. Chem.*, vol. 269, no. 12, pp. 9278–9283, Mar. 1994.

[5] P. J. George, C. Kuruvilla Jacob, Eds., *Natural rubber: agromanagement and crop processing*. Kottayam, India: Rubber Research Institute of India, 2000.

[6] G. F. J. Moir, *Nat. Lond. 184 1959 1926–1928*.

[7] A. S. Cook, B. C. Sekhar, 'Fractions from Hevea Brasiliensis Latex Centrifuged at 59,000 g', *Rubber Chem. Technol.*, vol. 27, no. 1, pp. 297–301, Mar. 1954, doi: 10.5254/1.3543482.

[8] S. H. J. B. Gomez, 'Frey-Wyssling complex in Hevea latex – uniqueness of the organelle'.

[9] P. Danwanichakul, W. Pohom, J. Yingsampancharoen, 'l-Quebrachitol from acidic serum obtained after rubber coagulation of skim natural rubber latex', *Ind. Crops Prod.*, vol. 137, pp. 157–161, Oct. 2019, doi: 10.1016/j.indcrop.2019.04.072.

[10] T. D. Pendle, P. E. Swinyard, 'The Particle Size of Natural Rubber Latex Concentrates by Photon Correlation Spectroscopy', *J. Nat. Rubber Res.*, vol. 6(1).

[11] J. Rafał Kędzia, A. Maria Sitko, J. Tadeusz Haponiuk, J. Kucińska Lipka, 'Natural Rubber Latex – Origin, Specification and Application', in *Application and Characterization of Rubber Materials*, G. Akın Evingür, Ö. Pekcan, Eds., IntechOpen, 2023. doi: 10.5772/intechopen.107985.

[12] D. C. Blackley, *Polymer latices. 1: Fundamental principles*, 2.ed. London: Chapman & Hall, 1997.

[13] K. Nawamawat, J. T. Sakdapipanich, C. C. Ho, Y. Ma, J. Song, J. G. Vancso, 'Surface nanostructure of Hevea brasiliensis natural rubber latex particles', *Colloids Surf. Physicochem. Eng. Asp.*, vol. 390, no. 1–3, pp. 157–166, Oct. 2011, doi: 10.1016/j.colsurfa.2011.09.021.

[14] S. Liao, 'A Review on Characterization of Molecular Structure of Natural Rubber', *MOJ Polym. Sci.*, vol. 1, no. 6, Dec. 2017, doi: 10.15406/mojps.2017.01.00032.

[15] J. T., P. Rojruthai, 'Molecular Structure of Natural Rubber and Its Characteristics Based on Recent Evidence', in *Biotechnology – Molecular Studies and Novel Applications for Improved Quality of Human Life*, R. Sammour, Ed., InTech, 2012. doi: 10.5772/29820.

[16] L. Tarachiwin, J. T. Sakdapipanich, Y. Tanaka, 'Relationship between Particle Size and Molecular Weight of Rubber from Hevea Brasiliensis', *Rubber Chem. Technol.*, vol. 78, no. 4, pp. 694–704, Sep. 2005, doi: 10.5254/1.3547907.

[17] S. Kovuttikulrangsie, J. T. Sakdapipanich, 'The molecular weight (MW) and molecular weight distribution (MWD) of NR from different age and clone Hevea trees', 2005.

[18] J. Rafal Kędzia, A. Maria Sitko, J. Tadeusz Haponiuk, J. Kucińska Lipka, 'Natural Rubber Latex – Origin, Specification and Application', in *Application and Characterization of Rubber Materials*, G. Akın Evingür, Ö. Pekcan, Eds., IntechOpen, 2023. doi: 10.5772/intechopen.107985.

[19] A. Subramaniam, 'Gel Permeation Chromatography of Natural Rubber', *Rubber Chem. Technol.*, vol. 45, no. 1, pp. 346–358, Jan. 1972, doi: 10.5254/1.3544711.

[20] Widyarani, S. C. W. Coulen, J. P. M. Sanders, M. E. Bruins, 'Valorisation of Proteins from Rubber Tree', *Waste Biomass Valorization*, vol. 8, no. 4, pp. 1027–1041, Jun. 2017, doi: 10.1007/s12649-016-9688-9.

[21] A. Akasawa, L.-S. Hsieh, Y. Lin, 'Serum reactivities to latex proteins (Hevea brasiliensis)', *J. Allergy Clin. Immunol.*, vol. 95, no. 6, pp. 1196–1205, Jun. 1995, doi: 10.1016/S0091-6749(95)70076-5.

[22] H. Y. Yeang, S. A. M. Arif, F. Yusof, E. Sunderasan, 'Allergenic proteins of natural rubber latex', *Methods*, vol. 27, no. 1, pp. 32–45, May 2002, doi: 10.1016/S1046-2023(02)00049-X.

[23] K. Berthelot, S. Lecomte, Y. Estevez, F. Peruch, 'Hevea brasiliensis REF (Hev b 1) and SRPP (Hev b 3): An overview on rubber particle proteins', *Biochimie*, vol. 106, pp. 1–9, Nov. 2014, doi: 10.1016/j.biochi.2014.07.002.

[24] N. Varghese, S. Varghese, S. Thomas, 'Radiation Processing of Natural Rubber Latex', in *Applications of High Energy Radiations*, S. R. Chowdhury, Ed., in Materials Horizons: From Nature to Nanomaterials., Singapore: Springer Nature Singapore, 2023, pp. 279–315. doi: 10.1007/978-981-19-9048-9_9.

[25] D. F. Parra, C. F. Pinto Martins, C. H. D. Collantes, A. B. Lugao, 'Extractable proteins from field radiation vulcanized natural rubber latex', *Nucl. Instrum. Methods Phys. Res. Sect. B Beam Interact. Mater. At.*, vol. 236, no. 1, pp. 508–512, Jul. 2005, doi: 10.1016/j.nimb.2005.04.028.

[26] K. P. Ng, E. Yip, K. L. Mok, 'Production of Natural Rubber Latex Gloves with Low Extractable Protein Content: Some Practical Recommendations', *J. Nat. Rubber Res.*, vol. 9(2).

[27] H. M. Ghazaly, 'Factory Production of Examination Gloves from Low Protein Latex'.

[28] A. Abu Hassan, n N. Abd Rahma, N. Abdullah, R. Sajari, M. K. Lang, 'Application of Proteolytic Enzyme in High Ammoniated Natural Rubber Latex', *Malays. J. Anal. Sci.*, vol. 22, no. 2, Apr. 2018, doi: 10.17576/mjas-2018-2202-14.

[29] K. Mariamma George, G. Rajammal, Geethakumary Amma, N. M. Mathew, 'Preparation and properties of low protein NR latex using stabilized liquid papain', *J. Rubber Res.*, vol. 10(3).

[30] N. M. V. Kalyani Liyanage, 'Colloidal Stability of Natural Rubber Latex', *Bull. Rubber Res. Inst. Sri Lanka*.

[31] L. P. Fah, 'Viscosity of Latex Mixes Using a Full Two-level', *J. Rubber Res.*, no. 6(4).

[32] K. Wadeesirisak *et al.*, 'Rubber particle proteins REF1 and SRPP1 interact differently with native lipids extracted from Hevea brasiliensis latex', *Biochim. Biophys. Acta BBA – Biomembr.*, vol. 1859, no. 2, pp. 201–210, Feb. 2017, doi: 10.1016/j.bbamem.2016.11.010.

[33] C. K. John, 'Coagulation of Hevea Latex with Surfactant and Salt', *J. Rubber Res. Inst. Malaya*, vol. 23, no. 2, pp. 147–156, 1971.

[34] J. C. Rodríguez Urbina, T. A. Osswald, J. E. Estela Garcia, A. J. Román, 'Environmentally safe preservation and stabilization of natural rubber latex in an acidic environment', *SPE Polym.*, vol. 4, no. 3, pp. 93–104, Jul. 2023, doi: 10.1002/pls2.10089.

[35] J. W. Van Dalfsen, 'The Effect of Rapid Stirring on Latex and on Its Creaming', *Rubber Chem. Technol.*, vol. 14, no. 2, pp. 315–322, Jun. 1941, doi: 10.5254/1.3540027.

[36] R. Alex, K. K. Sasidharan, T. Kurian, A. K. Chandra, 'Carbon black master batch from fresh natural rubber latex', *Plast. Rubber Compos.*, vol. 40, no. 8, pp. 420–424, Oct. 2011, doi: 10.1179/1743289810Y. 0000000038.

[37] 'Rubber Statistical News', *Statistics & Planning Department Rubber Board*, vol. 80, no. 12, May 2022.

[38] B. Kuriakose, 'Primary Processing', in *Developments in Crop Science*, vol. 23, Elsevier, 1992, pp. 370–398. doi: 10.1016/B978-0-444-88329-2.50023-4.

[39] T. C. Dickenson, *Filters and filtration handbook*, 3rd ed. Oxford, U.K: Elsevier Advanced Technology, 1992.

[40] N. Payungwong, J. Wu, J. Sakdapipanich, 'Unlocking the potential of natural rubber: A review of rubber particle sizes and their impact on properties', *Polymer*, vol. 308, p. 127419, Aug. 2024, doi: 10.1016/j.polymer.2024.127419.

[41] K. M. George, R. Alex, S. Joseph, K. T. Thomas, 'Characterization of enzyme-deproteinized skim rubber', *J. Appl. Polym. Sci.*, vol. 114, no. 5, pp. 3319–3324, Dec. 2009, doi: 10.1002/app.30642.

[42] 'IS 4588 (1986): Rubber, Raw, Natural'.

[43] R. Alex, C. Nah, 'Preparation and characterization of organoclay-rubber nanocomposites via a new route with skim natural rubber latex', *J. Appl. Polym. Sci.*, vol. 102, no. 4, pp. 3277–3285, Nov. 2006, doi: 10.1002/app.24738.

[44] D. Maga, V. Aryan, J. Blömer, 'A comparative life cycle assessment of tire recycling using pyrolysis compared with conventional end-of-life pathways', *Resour. Conserv. Recycl.*, vol. 199, p. 107255, Dec. 2023, doi: 10.1016/j.resconrec.2023.107255.

[45] A. Corti, 'End life tires: Alternative final disposal processes compared by LCA', *Energy*, vol. 29, no. 12–15, pp. 2089–2108, Dec. 2004, doi: 10.1016/j.energy.2004.03.014.

2
Rheology in Rubber Processing

Rosamma Alex, Pradeepkumar P. Joy, Sasidharan K. K.

2.1 Introduction

Rubber is a viscoelastic material and the elastic and viscous responses of pure raw rubber and raw rubber with additives play an important role in the processing of rubber products. The flow characteristics are affected by the various compounding ingredients, such as fillers, plasticizers and cross-linking chemicals. Maintaining suitable flow, good dispersion of different additives, obtaining the required level of vulcanization of rubber products and avoiding premature vulcanization are the main processing parameters that are significant in rubber product manufacturing.

By rheology of rubber is meant the study of the science of the flow of rubber (Greek: ῥέω – rheo, "flow" and λογία – logia, "study of"). Flow is a specific form of deformation and fluid flow is of particular interest, as it is a matter of daily experience. For example, it is generally observed that food materials, such as curd and butter, undergo a change of texture from a consolidated mass to a more fluid form when stirred. Heating sugar produces caramel, which is very viscous, whereas sugar dissolved in water exhibits much lower viscosity. Early studies on fluid flow were later extended to the materials exhibiting behavior intermediate between fluids and elastic solids. Such materials are called viscoelastic fluids. The viscosity of fluids and polymer solutions is determined with capillary viscometers, which determine flow characteristics based on the time required for a sample to flow through a capillary. A falling-ball viscometer measures the viscosity of a fluid based on the time a sphere of known density takes to pass through two points in the fluid. In parallel, instruments employing motor-driven spindles that have different shapes and sizes and rotate at defined speeds (revolutions per minute or rpm) in fluids were being developed. The torque required to rotate the spindle increases with fluid viscosity. In polymer solutions, viscosity correlates with molecular weight while, in polymer melts, it is influenced not only by molecular weight [1], but also by molecular weight distribution, entangle-ment and related parameters. Additionally, viscosity also depends on the shear rate, temperature and the presence of additives, especially fillers such as carbon black.

One of the instruments developed to study the viscous and elastic behavior of visco-elastic materials was a modified capillary viscometer [2]. An early study of the visco-elasticity of polymers was based on stress-relaxation experiments, which provided information on polymer relaxation times [3, 4]. Later, rotational viscometers and os-cillating rheometers were developed and they are currently used to determine the flow and deformation of polymers, specifically rubber.

These different operations for processing rubber to rubber products heavily depend on the flow or rheological aspects, as evident from published books on the rheology of rubber [5, 6, 7]. Rubber flows because of the shear forces developed in various mixing and shaping machines, which are affected by the additives. Commonly used mixing equipment are (a) two-roll mixers (open mills), (b) internal mixers with tangential rotors or intermeshing rotors and kneaders and (c) continuous mixers. The shaping operations are mainly done by extruders, calenders and hydraulic presses during molding.

Compression plastimeters measure the plasticity of rubber in terms of the raw rubber specimen's thickness under specified conditions of load and temperature. Rotational viscometers are used to study rubber flow in terms of viscosity, which is the torque required to rotate a rotor. The flow characteristics related to viscoelastic parameters can be studied by oscillating disk rheometers and moving-die rheometers. A capillary rheometer studies the extrusion characteristics of relevance to processing.

2.1.1 Deformation of Ideal Viscous Fluids and Ideal Elastic Solids

When a shear stress is applied to a fluid layer, the deformation is visualized as a flow of the fluid in a particular pattern and, in this type of deformation, no energy is stored. The fluid will flow as long as shear force is applied. The flow pattern formed is due to the low intermolecular force within the fluid.

Unlike a solid, when this shear force is removed, the fluid does not regain its original shape. When fluids begin to flow, the layers need to overcome the intermolecular forces and the energy used for this is dissipated by the fluid as heat. So, in an ideal viscous material, no energy is conserved during deformation and the material adopts the shape of its container – exhibiting Newtonian flow behavior.

In general, in an ideal elastic material, the strained shape is fully reversible and there is no energy loss during the deformation and recovery of the material. The elastic solids obey Hooke's law during deformation. In elastic solids, shear deformation is quantified by shear modulus. If a solid is deformed and the strain is kept constant, the elastic solid can remain at rest in the strained condition. During deformation, the elastic solids store energy, which is released as kinetic energy when they recover their original shape upon removal of the deforming force.

2.1.2 Stress and Strain in Solids

The deformation in a fluid or liquid is quantified as strain. Strain is the deformation experienced by solids or fluids and is expressed in length, thickness, volume or shape depending on the deformation stress applied. Strain is quantified as a ratio and is a dimensionless number. Strain may be classified as tensile, bulk (or volumetric) or shear strain, each corresponding to a specific direction of applied stress.

For most polymers, the relation between strain and stress is linear only to a limited strain level, above which stress–strain behavior is non-linear. Elastomers are polymers that undergo very high elongation and, when the strain is low, the stress is directly proportional to strain. The proportionality constant in this relation is called the elastic modulus.

Viscosity and types of fluid flow

When a shear stress is applied to a fluid surface, its various layers flow at different velocities on account of internal friction between the fluid layers, a phenomenon attributed to fluid viscosity. If a shear force is applied to the top layer, it moves at a higher velocity than the underlying layers. Figure 2.1 illustrates how the movement of the different fluid layers resembles the sliding of a deck of cards when the top card is pushed gently.

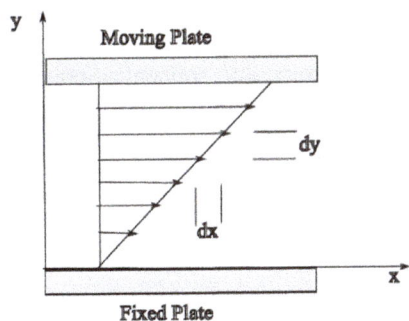

Figure 2.1 Schematic representation of fluid flow between two plates, illustrating the velocity profile of different fluid layers

Applying a shear force to a fluid can be better represented by considering the application of force to a plate on the surface of the fluid. The top layer of fluid moves with maximum velocity (u max.). The lower layers will not move at the same speed as the top layer, as a frictional force called viscous force opposes the movement of the fluid layer just below. For the successive lower layers, flow occurs at a lower and lower speed. The layer in contact with the lower stationary plate can be assumed to have zero velocity or to be in a no-slip condition. This is because of the attraction between molecules of fluid and atoms on the surface of the stationary plate. Hence, during flow, there will be a velocity profile for the fluid (Figure 2.1).

Based on the velocity profile, the velocity difference between two adjacent layers (du) can be expressed in terms of the difference in distance between the two layers (dx) over a certain time (dt):

$$du = dx/dt \tag{2.1}$$

The deformation experienced by fluids is quantified as shear strain γ, which is defined as γ = lateral deformation/height of the layer being deformed:

$$\gamma = dx/dy \tag{2.2}$$

The shear strain rate is the velocity gradient within the different fluid layers. Taking the depth difference between two adjacent layers to be dy:

$$\dot{\gamma} = \text{shear strain}(\gamma)/\text{time taken}(dt) \tag{2.3}$$

$$\dot{\gamma} = dx/dt.dy \tag{2.4}$$

$$\dot{\gamma} = du/dy \tag{2.5}$$

The shear stress (τ) is proportional to the shear strain rate for many fluids, such as water and oil.

$$\tau = \text{constant} \times \dot{\gamma} \tag{2.6}$$

$$\tau = \eta\,\dot{\gamma} \tag{2.7}$$

η is called viscosity. Viscosity is affected by the temperature and concentration of a fluid.

Polymer flow can be classified as Newtonian and non-Newtonian.

2.1.3 Newtonian Flow

Fluids in which shear stress is directly proportional to shear strain rate are called Newtonian fluids. On a graph, shear stress is zero when the shear strain rate is zero. According to Newtonian flow behavior, fluid flow is faster when the shear force is higher.

A liquid such as honey (fluid of high viscosity) flows more slowly than a fluid such as water (fluid of low viscosity) when the same shear stress is applied to both fluids. Honey and water are considered Newtonian fluids.

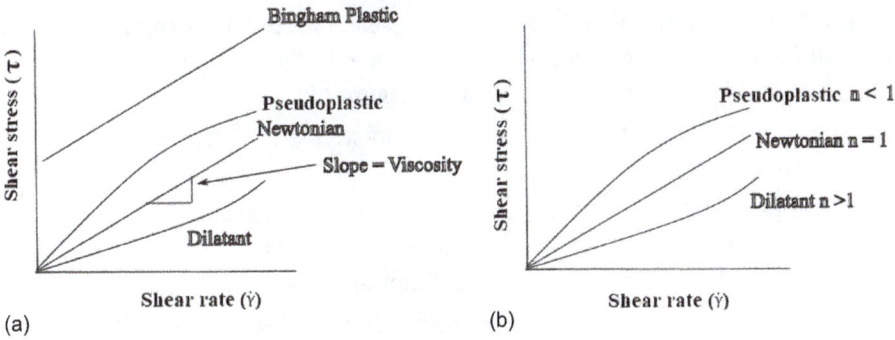

Figure 2.2 (a) Schematic representation of Newtonian and non-Newtonian flow, (b) power-law index for Newtonian and non-Newtonian flow

2.1.4 Non-Newtonian Flow

Non-Newtonian fluids do not always have constant viscosity at different shear stress and shear rates. The flow other than Newtonian can mainly be based on variation of shear stress and shear rate, which is not linear. Pictorial representations of the behavior of Newtonian and non-Newtonian flow are shown in Figure 2.2 (a).

Pseudoplastic (shear-thinning) flow

In certain fluids, shear stress varies directly with the shear rate at very low shear stress values but, at the higher shear rate, the shear stress changes only marginally. In technical terms, this behavior can be called shear thinning and the flow is described as pseudoplastic.

In this, the most common type of non-Newtonian behavior, the fluid viscosity decreases with increase in shear rate. Such fluids show an almost constant shear viscosity (η_0) called zero shear viscosity. Shear thinning is marked by a slight reduction in viscosity beyond a critical shear rate, with the shear rate–shear stress behavior above this point defining the shear-thinning region.

In the shear-thinning region, the shear stress–shear rate variation is expressed as:

$$\tau = k \, \dot{\gamma}_n \tag{2.8}$$

Where

n = power-law index and is related to the extent of thinning

k = flow consistency index

$$\log \tau = \log k + n \log \dot{\gamma} \tag{2.9}$$

k and n can be obtained from a graph of log σ versus log $\dot{\gamma}$. Polymer melts and rubber exhibit pseudoplastic flow. According to the power-law index, flow types are categorized as follows: for Newtonian fluids, $n = 1$; for pseudoplastic (shear-thinning) fluids, $n < 1$; and for dilatant (shear-thickening) fluids, $n > 1$. The response of viscosity to varying shear rates for these fluids is shown in Figure 2.2.

Pseudoplastic fluids exhibit a decrease in viscosity at higher shear rates, as may be seen in Figure 2.3.

Examples of pseudoplastic materials are rubber, tomato sauce and curd. A practical observation in the flow of these materials is that, when they are sheared well, their viscosity decreases. An advantage of shear-thinning behavior is observed in paint, which remains sufficiently viscous to be transferred into small containers and be picked up by a brush. However, when it is applied to a surface, it experiences a higher shear rate, due to the rapid back-and-forth movements, and this causes the paint to become less viscous.

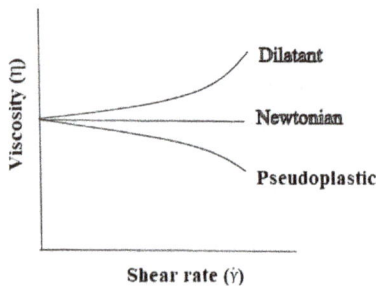

Figure 2.3 Variation in viscosity with shear rate for different types of flow

2.1.4.1 Dilatant (Shear-Thickening) Flow

Dilatant fluids show a sudden increase in viscosity at certain high shear rates. Such flow can be described as shear thickening. This phenomenon is also commonly referred to as dilatancy.

The flow of corn starch is dilatant. While mixing corn starch with water, we observe that it becomes thicker when stirred rapidly at high shear rates, but remains less viscous under slow mixing..

2.1.4.2 Bingham Plastic

In some fluids, no flow occurs until a certain shear stress level is applied and such fluids are called Bingham plastic fluids. Certain types of pastes, certain medical creams and toothpaste exhibit this behavior. For instance, the paste will not flow if we hold the tube upside down. However, when squeezed with force (significantly increasing the stress), a larger quantity of paste comes out, as the viscosity of the paste decreases.

2.1.5 Time-Dependent Viscosity

The variation in the viscosity of a fluid with shear stress depends on the duration of the shear stress applied. So, at a constant shear rate, viscosity may increase or decrease with time. This effect is described as being rheopectic or thixotropic. Systems for which the relation between viscosity and shear rate at constant shear stress or the relation between viscosity and shear stress at constant shear rate exhibit a dependence on the duration of shearing or application of stress are called time-dependent fluids.

Based on the time-dependent viscosity variation, fluids can be classified as thixotropic and rheopectic. A thixotropic fluid sheared at a constant rate ($\dot{\gamma}$) exhibits a time-dependent decrease in apparent viscosity, accompanied by a corresponding reduction in shear stress, as described by the relation $\eta = \sigma / \dot{\gamma}$.

2.1.5.1 Thixotropic Fluids

Fluids in which the viscosity at constant shear rate decreases with time are called thixotropic fluids, as shown in Figure 2.4 (a). For most liquids, the variation in viscosity with shear stress is reversible because, if the shear stress is removed, the fluid recovers almost all its original viscosity. For thixotropic fluids, this recovery happens after a specific length of time. For pseudoplastic fluids, there may be a time-dependent rearrangement in the molecular structure of the fluid at higher shear stress. So, a pseudoplastic fluid can also exhibit thixotropic behavior. Not all shear-thinning materials exhibit time-dependent viscosity changes. But generally, a thixotropic material will be shear-thinning.

A thick mixture of sand mixed and water (called quicksand) has a thixotropic flow behaviour. An example of a material's thixotropic nature can be observed by walking quickly on quicksand. The higher shear force applied by the rapid movement causes the sand to behave more like a solid. Conversely, standing still exerts less shear force and the legs will sink.

The time-dependent flow pattern of a thixotropic fluid can be better understood by observing the shear-rate–shear-stress variation during a loading and unloading cycle. The curve formed by the loading and unloading cycle has a concave shape during the loading cycle only. Thixotropy is important in applications such as paints, coatings and surface leveling with polymeric materials. Thixotropic latex paints, such as styrene–acrylate-based emulsions, experience a decrease in viscosity when applied with a brush or roller, and this allows easy spreading. Once brushing stops (removing the shear), the viscosity of the paint rapidly rises again, thereby preventing dripping and sagging.

2.1.5.2 Rheopectic Fluids

Fluids in which the viscosity at a constant shear rate increases over time are called rheopectic fluids (Figure 2.4). Many fluids, such as printing ink, silly putty (silicone-based polymer), certain lubricants and protein solutions, when maintained at a constant shear rate for a longer time, have a tendency to solidify, because the viscosity increases. When the silly putty is squeezed/exposed to shear forces, it becomes more viscous. When the deforming force is removed, it regains its original lower viscosity. The rheopectic behavior of a rheopectic fluid (silly putty) is shown schematically in Figure 2.4.

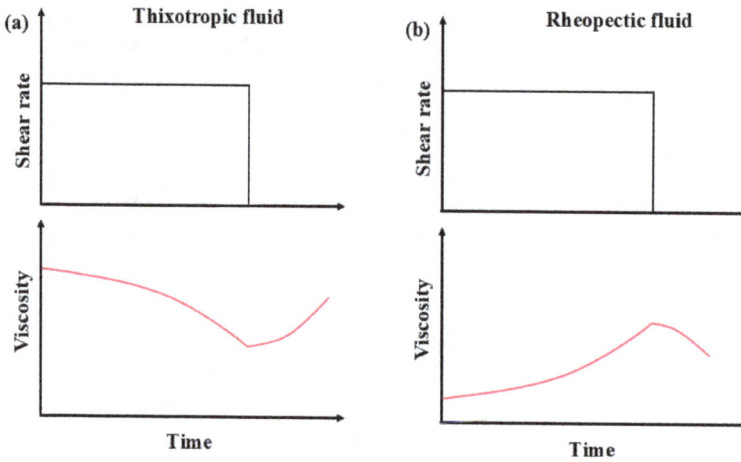

Figure 2.4 Change in viscosity with time for (a) a thixotropic fluid and (b) a rheopectic fluid

2.2 Rubber as a Viscoelastic Material

Polymers, in general, are viscoelastic materials; however, their viscous and elastic natures vary. The flow behavior of polymer melts is generally non-Newtonian. The viscosity or fluidity (fluid-like nature) of rubber is higher in the unvulcanized state, making the raw rubber easily deformable under external forces. Raw rubber exhibits significantly lower elasticity than vulcanized rubber. A mathematical model of a spring (elastic component) and dashpot (viscous component) arranged in series can be used to understand the viscoelasticity of raw rubber. In this arrangement, the dashpot and the spring show the same stress when the system is subjected to an external force, while the strain realized in the dashpot is different from that in the spring. Since the dashpot is free to deform, the viscoelasticity in raw rubber is observed as time-dependent viscous flow, with the spring offering resistance to flow during fast deformations.

Cross-linked rubbers have a more elastic nature and are classed as viscoelastic solids. Their combined viscous and elastic natures give rise to time-dependent stress–strain behavior, making their deformation only partially reversible. As a result of time-dependent deformation, stress decays under constant strain while strain increases under constant stress.

Raw rubber's non-Newtonian flow behavior is attributed to the very high molecular weight of raw rubber, which is entangled and has a branched network structure. Rearrangement in the rubber network arising from conformational changes can lead to shear-thinning behavior during rubber flow.

Shear thinning of rubber and polymer plays a key role in polymer processing.

The rubber flow is affected by factors such as temperature, shear rate, shear stress, viscosity microstructure and macrostructure of rubber, the presence of additives and the method of incorporating additives. Rubbers are materials with very low glass transition temperatures and are flexible at ambient temperatures. Flexibility increases as temperature increases above ambient.

2.2.1 The Flow of Rubber and Rubber Compounds during Processing to Products

Rubber is a material that is used as a product after it has been compounded and cured or vulcanized.

Rubber compounds consist of several categories of ingredients, including elastomers, vulcanizing chemicals, fillers, anti-degradants, plasticizers and specialized additives. These influence the rubber's physical and chemical properties. Physical properties, including the viscous and elastic natures of the rubber compound, play a critical role in specific processing steps, such as mixing, shaping and vulcanization. The uncured properties governing flow and the cured properties resulting from cross-linking are essential for maintaining product quality. The chemical properties of rubber are closely linked to the aging resistance of the compounds, which is also essential for maintaining the quality of rubber products.

Rubber products contain several additives that play a critical role in determining the rheological behavior of the rubber. These additives significantly influence the rheology of rubber and the mechanical properties achieved after vulcanization. The presence of fillers, specifically reinforcing fillers, impacts the flow of rubber compounds in terms of surface finish, shrinkage and die swell [7].

The fillers affect the viscoelasticity of rubber. Carbon black interactions may lead to the formation of agglomerates through physical bonds. These physical bonds can break and reform during deformation, resulting in energy dissipation. Good filler dispersion that prevents this physical bond formation can lead to low energy losses during deformation. These factors can affect the die swell observed in extrudates.

2.2.1.1 Elastomer

The selection of a specific rubber is based on the product's property requirement and service life. Natural rubber or synthetic rubbers, alone or in blends, serve as the base elastomeric matrix. Chemically, these are saturated and unsaturated rubbers or polar and non-polar rubbers while, from an applications viewpoint, they are general-purpose and special-purpose rubbers. Polybutadiene rubber (BR), styrene–butadiene rubber (SBR), polyisoprene rubber (IR) and ethylene propylene rubber (EPM & EPDM) are considered the main general-purpose rubbers. These have comparatively very good mechanical and elastic properties. So, most of them are used in general rubber goods, including tires.

Special-purpose rubbers have at least one particular property along with basic elastic properties, e.g. resistance to aging, resistance to chemicals, resistance to swelling in non-polar oils, resistance to high or low temperatures, etc. A few special-purpose rubbers are acrylonitrile–butadiene rubbers (NBR), butyl rubber (IIR), polychloroprene rubber (CR), silicone rubbers, fluorocarbon rubber and polyurethane rubber. BR and SBR are extensively used in the tire sector, specifically for passenger vehicles. Butyl rubber (IIR) has excellent air impermeability characteristics and is used in tire tubes. Halobutyl rubber, *viz.*, chlorobutyl and bromobutyl rubber find application in tubeless tires as inner liners that act as tire tubes. EPM and EPDM have excellent ozone resistance and heat resistance. NBR is noted for its superior oil resistance. Silicone rubber is considered to be non-toxic and has applications in body implants, pharmaceuticals and food-contact products. As a shock absorber, it is used for both low and high temperatures and it has moderate resistance to oils and excellent heat and ozone resistance. Fluorocarbon rubbers have excellent heat, ozone and fire resistance. Different synthetic rubbers and natural rubber have different chemical resistance and compatibility levels. The chemicals used in rubber processing need to be selected based on this factor.

Several modified forms of natural rubber are used in rubber processing for specific applications. Superior processing (SP) rubber consists of a blend of vulcanized and unvulcanized natural rubber obtained by intimately mixing natural and pre-vulcanized latex in specific ratios. Processing aid (PA) rubber contains a non-staining oil dispersed in SP rubber. Superior processing rubber, which consists of 20 parts vulcanized rubber and 80 parts unvulcanized rubber, is called SP 20. It is available as crepe or as sheet. SP is available as masterbatches that they can be mixed with virgin rubber to achieve the required content of vulcanized rubber [8, 9]. For example, processing aid 80 contains 80% vulcanized rubber. PA 57 is obtained by mixing 80 parts of cross-linked rubber and 20 parts of unvulcanized rubber in the latex stage. Before latex coagulation, a stable emulsion of light-colored, non-staining oil is added to yield a concentration of 40 parts in PA 57.

SP rubber or PA rubber is useful for reducing die swell in extruded products and obtaining dimensional stability in calendered and extruded products.

2.2.1.2 Fillers

These are mainly added to enhance the properties of rubber. Adding filler boosts the volume of rubber, leading to a cost reduction for the product, without adverse effects. The filler reinforces the rubber matrix by entering into specific physical or chemical interactions with it. The three-dimensional existence of a filler, such as carbon black or silica, with voids can be described as a structure, since the particles can form agglomerates. In raw rubber compounds, filler–rubber interactions, both physical and chemical, and the presence of structure can be related to viscosity change and flow characteristics.

The mixing of reinforcing fillers with rubber by means of the shear forces produced in the mixing equipment, such as an open mill or internal mixer, is an essential processing step for the effective interaction of filler with rubber and reinforcement.

In filler-loaded rubber compounds, the presence of rigid filler particles induces a hydrodynamic effect that reduces the velocity gradient during flow and results in increased viscosity. Additionally, weak interactions between the rubber matrix and filler can further elevate viscosity. These factors reduce the flow of rubber and lead to strain amplification, whereby the strain in the rubber is increased by the presence of filler.

Unlike unfilled rubber, rubber reinforced with filler exhibits more non-Newtonian behavior, as revealed by several studies related to the flow of carbon black-incorporated rubber [10, 11]. These studies noted a strong shear-thinning behavior with reinforcing carbon black. It was also observed that there was an increase in viscosity as the loading of carbon black increased and as the surface area of carbon black increased. Increasing the structure of carbon black generally leads to lower extrusion shrinkage, smoother extrusion, reduced shear-thinning behavior (as indicated by a higher power-law index, n), and increased compound modulus and viscosity.

2.2.1.3 Textile Materials in the Rubber Industry

Different textiles and steel wires are used in the rubber industry for hoses, transmission belts, conveyor belts and tires. Cotton (from natural cellulose) was the first textile material used but, by the end of the 19th century, a synthetic fiber, rayon based on cellulose, was developed. Cotton is now mostly replaced by synthetic types of textiles. Rayon is a regenerated cellulose fiber widely used in tire and industrial rubber products. Fully synthetic fiber nylon 66, a polyamide, was developed by DuPont and was commercialized in 1936. Different types of polyamides (marketed as nylon by DuPont) were developed later. Nylon has significantly superior tensile strength and fatigue resistance compared with cotton and rayon.

Polyester, introduced in 1942, retains its importance as a standard reinforcing textile material in rubber products, mainly in the tire industry. Polyester fiber has a high modulus and tensile strength and is the most popular reinforcing material for tires. The disadvantage is that it has poor adhesion qualities and undergoes thermal shrinkage. Specific fabric treatments are needed for good adhesion to rubber and its application in tires.

Polyaramid (marketed as Kevlar by DuPont) has a very high tensile strength, rigidity and low elongation at break. In terms of strength, it is equal to steel, with the advantage of being lower in density. However, its adhesion to rubber is not good.

Glass fiber has high tensile strength and rigidity, resists high temperatures and has poor bending and fatigue strength. Steel wire in the form of steel cord is used widely in the belts of radial passenger tires. It can also be used in the carcasses of truck tires. Steel has a high tensile strength, rigidity and good resistance to high temperatures. In addition to its use in tires, it is employed for hoses, driving belts and conveyor belts.

Synthetic fibers, such as rayon, nylon and polyester, are coated with suitable adhesive solutions and dried to enhance bonding with rubber during vulcanization.

One such adhesive is resorcinol formaldehyde latex (RFL). Its composition is that of a rubber latex mixed in a solution of resorcinol and formaldehyde in water. For rayon (based on cellulose), SBR latex is sufficient; for nylon, a more polar latex, a blend of SBR and vinyl pyridine latex, is used.

In the first stage of adhesive preparation, an aqueous solution of resorcinol and formaldehyde is prepared, mixed with sodium hydroxide and kept for several hours at room temperature. In the second stage, the resin solution is added to the latex. The latex is dispersed in the continuous resin phase.

A typical formulation for resorcinol formaldehyde latex adhesive for fabric treatment [12] is given in Table 2.1. The fabric is dipped in an RFL-bonding system and dried. The treated fibers so obtained have improved adhesion to rubber. The ratio of resorcinol and formaldehyde can be 2:1.

Table 2.1 Typical Formulation for Resorcinol Formaldehyde Latex Adhesive

Ingredient	Parts by Weight
First stage – Preparation of resin solution	
Resorcinol	13.0
Formaldehyde, 37%	18.9
Sodium hydroxide, 10%	8
Water	306.5
Second stage – Preparation of latex component	
SBR latex, 40%	50
Styrene–vinyl pyridine–butadiene terpolymer latex	200
Water	19.3

The RFL adhesive is prepared by mixing resin and latex

2.2.1.4 Processing Aids

Plasticizers and softeners are used in rubber compounding mainly to aid the dispersion of compounding ingredients, such as fillers, because good filler dispersion enhances mechanical properties. They also help to reduce viscosity to an extent depending on the quantity of plasticizer added, which is significant in rubber processing. They adversely affect processing characteristics and mechanical properties when used at higher levels.

The main factor determining the compatibility of rubber matrix and plasticizers is the degree of chemical similarity, based on whether the rubber and plasticizer are polar or non-polar. Usually, polar rubbers are compatible with polar plasticizers. Naphthenic oil is a commonly employed plasticizer. A polar plasticizer, such as dioctyl phthalate (DOP) or dibutyl phthalate (DBP), is used for polar rubber, such as nitrile rubber. In contrast, a non-polar rubber, such as EPDM paraffin oil is preferred as the processing aid. Pine tars are highly compatible with natural rubber, provide good filler dispersion and improve compound properties, such as adhesion and fatigue resistance. Hydrocarbon oils are the most commonly used processing aids. Other important plasticizers are vegetable oils, fatty acid salts and rosin. They are useful for diluting rubber matrix, thereby enabling more filler addition and cost reduction.

The plasticizers are helpful during processing operations, such as molding, calendering, coating and extrusion, where they increase the mobility of the polymer chains and lower the viscosity without chemical interaction with rubber.

The shear rates required for these processing operations are given in Table 2.2. The viscosity and flow parameters change upon addition of compounding ingredients, such as fillers and plasticizers. Generally, to ensure good dispersion of non-reinforcing or semi-reinforcing fillers, the plasticizer added is $1/20^{th}$ of the filler. However, for carbon black filler, the plasticizer added is $1/10^{th}$ of the filler. For calendering operations, generally, the plasticizer used is higher and about $1/5^{th}$ of the filler content. Incorporating more plasticizers can also benefit extrusion – approximately $1/10^{th}$–$1/8^{th}$ of the filler.

Table 2.2 Shear Rates Required in Various Processing Operations of Polymers

Processing	Shear Rate (s^{-1})	Examples of Rubber Products
Compression molding	1 to 10	Molded rubber products, such as gaskets, footwear and o-rings
Calendering	10 to 100	Hospital sheetings and general beltings and coated fabrics, tire carcass materials
Extrusion	100–1000	Hoses, tubes, tire tread
Injection molding	10^3 to 10^4	Pharmaceutical closures, such as injection bottle caps

2.2.1.5 Anti-Degradants

These are used to protect rubber during processing and in service. Time, temperature, ozone, light and oxygen affect rubber by chemical changes collectively known as aging. Synthetic rubbers offer better aging resistance than natural rubber. Anti-degradants are added to the compound to avoid polymer aging and to extend its service life.

2.2.1.6 Vulcanizing Chemicals

Vulcanization is a process in which chemical cross-links are formed between polymer chains, changing raw rubber from a plastic to a highly elastic form. All the mechanical and dynamic properties improve after vulcanization. Hence, the process of vulcanization controls the final rubber properties. Elemental sulfur is the most widely used vulcanizing agent, but peroxides and metal oxides are used to an extent depending on the rubber. Sulfur and sulfur-bearing chemicals are used for NR, SBR, IIR, BR, EPDM and NBR. Organic peroxides serve as the vulcanizing agent for saturated rubbers, such as EPM and silicone rubber. Co-agents such as triallyl cyanurate (TAC) and triallyl isocyanurate (TAIC) are used along with peroxides to raise the state of vulcanization. Other vulcanizing agents are metal oxide for halogen-containing rubbers, such as CR and CSM. Dioximes and phenolic resins serve as vulcanizing agents for specific grades of butyl rubber.

Rubber can also be vulcanized without the addition of external vulcanizing agents through interaction with functional groups of suitably functionalized rubbers. Functionally modified rubbers, such as epoxidized natural rubber chlorosulfonated polyethylene and polychloroprene rubber, can undergo cross-linking involving their functional groups. Such rubber blends are called self-vulcanizate rubber blends [13, 14].

The vulcanizing agents are selected on the basis of such factors as the chemical reactivity of the rubber, the type and number of cross-links required and the stability of the cross-links. These factors related to rubber cross-linking significantly affect a rubber product's performance.

In general, cross-link density increases with cure time. Chemicals that help to raise the vulcanization rate and enhance the degree of vulcanization are called accelerators and activators. Thiazoles, sulfinamides, dithiocarbamates and guianidines are common accelerators while ZnO and stearic acid are activators. The activators and accelerators form complexes that activate the curing process.

2.2.1.7 Special-Purpose Additives

These chemicals are added to impart specific properties and are not used in general-purpose rubber products. Some special-purpose additives are blowing agents (for expanded rubber), flame retardants (for cables and specific rubber products), pigments and colors (for toys, play balls, balloons, hot water bags, erasers, etc.), anti-static agents (anti-static gloves) and abrasives (for unique mats and special erasers).

2.2.2 Rubber Formulation

Rubber formulation involves drawing up a list of the compounding ingredients and their quantities relative to the amount of elastomer. Usually, in a recipe, each ingredient is expressed as parts per hundred parts of rubber (phr), where the amount of elastomer is taken as 100 parts. Each particular compound has its own recipe.

2.3 Processing of Rubber

Rubber is a viscoelastic material and a related property, such as modulus, is susceptible to temperature changes. As the temperature of rubber increases from below room temperature, four distinct phases can be observed:

1. Glassy region

2. Glass transition (T_g) region

3. Rubber region

4. Rubber-flow region

As the name suggests, in the glassy region, rubber behaves like a glass at very low temperatures. Natural rubber becomes very stiff, with almost no elastic properties at temperatures below −70 °C, the T_g of natural rubber. In the glass transition region, the rubber molecules slowly acquire the ability to rotate. Consequently, the modulus decreases gradually and, by the end of the glass transition region, it declines by several decades. As the temperature rises above room temperature, the rotational and segmental motion of the rubber molecules increases, causing the polymer to become more rubbery and exhibit enhanced elastic behavior. Again, as the temperature rises, the rubber region changes to the rubber-flow region. Rubber is processed in the rubber-flow region.

2.3.1 Steps Involved in the Processing of Rubber

The first step in product manufacture is the mixing of elastomers and additives in suitable mixing equipment. In this process, the additives are selected for their ability to achieve the required properties during the shaping operations and meet the desired performance of the product. The rubber compound thus obtained is shaped by different methods, such as calendering, extrusion, molding techniques (e.g., compression and injection molding) and coating. After shaping, the rubber product is vulcanized to obtain suitable mechanical properties and service requirements. Vulcanization may occur during shaping, as happens in molding techniques. In other methods, vulcanization occurs after the compound has been shaped.

A high viscosity is needed during the incorporation of additives, mainly fillers. The reinforcement provided by fillers happens as the shear forces are being exerted during mixing on the mixing mill and, hence, a higher viscosity is preferred. Generally, a Mooney viscosity of 45 units yields a smooth band on a mixing mill to which compounding ingredients can be added. The minimum initial plasticity (P_0) for technically specified rubber is set at 30 Wallace plasticity units. Plasticity is directly related to the material's viscosity. For the shaping operations, lower viscosity is required for the sake of lower energy consumption and ensuring smooth, uniform flow.

The flow of rubber is affected by the shear forces developed in the various mixing and shaping machines. The commonly used mixing equipment are (a) two-roll mixers (open mills), (b) internal mixers with tangential rotors or intermeshing rotors and kneaders and (c) continuous mixers.

Factors affecting mixing on two-roll mills include (a) the distance between the two rolls, called the nip gap; (b) the compound that remains above the nip, called the rolling bank, and (c) the thickness of the compound wrapped around the roll, called the band. These elements affect mixing efficiency by controlling shear forces generated as the rubber and additives pass through the rolls. The friction ratio, defined by the difference in speed between the front and back rolls, also plays a key role. Generally, the front roll rotates three times while the back roll rotates four times to enable good rubber compound cutting and maintain a good bank while the additives are being incorporated.

A standard laboratory mixing mill can be 13 inches x 6 inches (length times diameter of the rolls) and has the capacity to make 1–1.5 kg rubber compound. There are industrial mixing mills with larger rollers measuring 26 inches x 84 inches with a capacity of 80–140 kg and 30 x 100 inches with a 100–180 kg capacity.

The main advantage of a mixing mill is excellent mixing quality and no scorching. The disadvantages are longer mixing times, operator-dependent mixing quality and the spread of rubber chemicals around the mixing place.

The two types of internal mixers that are frequently used are the Banbury, also known as a tangential mixer, and the intermix, also known as an intermeshing mixer. In both these machines, mixing quality is controlled by the temperature and speed of the rotors. Tandem mixer technology is another transformative advancement for the tire industry, enabling a high-performance, two-stage mixing process that enhances the quality and efficiency of mixing.

Tangential mixers like Banbury are characterized by a flow of the compound from the sides of the mixing chamber toward the center, where the shear mixing occurs. As the rotors move at different speeds, mixing occurs by shearing the rubber and additives between the chamber wall and rotor. Banbury mixers offer shorter mixing cycles, higher machine efficiency and faster feeding and incorporation of the material.

In the intermix, the rotors have the same speed and mixing occurs mainly in the gap between the rotors by shearing between the intermeshing rotors. Generally, intermeshing rotors are bigger than tangential rotors and occupy a greater space. So, the fill factor is smaller than in the case of tangential rotors. The mixing quality is excellent. The flow properties of rubber are altered by adding reinforcing fillers; the changes occur primarily due to rubber–filler interactions during the mixing stage.

In tandem mixing, both the master batch and final batch are mixed consecutively within the same setup, combining two stages into one efficient process. This setup consists of a conventional internal mixer and a larger tandem mixer positioned directly below, without a floating weight. After initial mixing in the top unit, the master batch is discharged into the tandem mixer, which synchronizes with the first mixer. The masterbatch cools to 100–120 °C while mixing is completed. Meanwhile, the top mixer processes the next master batch. Once the tandem mixer cycle ends, the finished mix is discharged to a mill or extruder for further processing. The primary advantage of this method is that it eliminates the need for intermediate storage of the masterbatch between stages [15, 16].

It is preferable to have good rubber–filler interactions and fewer filler–filler interactions to allow uniform flow during processing. Good filler dispersion enhances the reinforcement ability of carbon black and promotes better entanglement of rubber chains, improving the overall mechanical properties of the rubber. There is better rubber–filler interaction if filler dispersion is good. The type of filler and the incorporation method strongly affect filler dispersion characteristics.

2.3.1.1 Mastication of Rubber

Mastication reduces the viscosity of rubber to a processable range, which helps with better filler dispersion. An internal mix or open mill masticates natural rubber by applying shear stress. The molecular chains that break under shear forces become stabilized by atmospheric oxygen. The molecular weight may be reduced to about 10^6 from 10^7 or the Mooney viscosity may be reduced to about 45 units. At low temperatures, breakage of polymer chains occurs exclusively by mechanical shear forces. However, at high temperatures, thermo-oxidative aging occurs, due to the formation of peroxides that can adversely affect the quality of rubber. During mastication, the rubber chains undergo scission and the atmospheric oxygen stabilizes the free radicals formed as rubber chain segments. As the temperature of mastication increases, a higher level of chain scission occurs. The greater mobility prevents shear forces from being exerted, because the chains slide past each other. Mastication is preferred at low temperatures and its efficiency can be increased with the aid of peptizers. Mixing is performed by circulating cold water during mastication.

2.3.1.2 Mixing Process on a Two-Roll Mill

The most challenging part of rubber mixing is dispersing the filler. This involves four different steps: incorporation, plasticization, dispersion and distribution.

Incorporation: In the incorporation phase, the flexible and entangled rubber chains undergo shear forces, embedding filler aggregates into the rubber matrix and forming a coherent mixture of additives. Initially, the polymer "wets" the filler and encapsulates fragments formed from filler and allows for further size reduction of filler fragments. Under increased pressure and shear in the mixer, the polymer fills gaps within the filler agglomerates, displacing any trapped air. This phase, known as incorporation or wetting, allows the initially separate components to merge into a uniform mass before the beginning of dispersion and distribution. Incorporation occurs through two main mechanisms: in one, the elastomer deforms extensively, increasing surface area and sealing in filler agglomerates; in the other, the elastomer breaks into smaller pieces, blends with the filler and encapsulates it once again [6, 15, 17].

Plasticization: During mixing, the polymer chains undergo chain scission, reducing the molecular weight. The level of mastication achieved for the purpose of enabling filler incorporation generally does not adversely affect the quality of rubber products.

Dispersion: In the dispersion phase that follows incorporation, shear stress breaks down filler agglomerates to their smallest possible size, distributing them throughout the rubber matrix. Initially, carbon black forms larger agglomerates ranging from 10–100 µm, which are gradually reduced to less than 1 µm during dispersion. This process further distributes the smaller agglomerates and aggregates derived from the filler pellet fragments. Usually, at the end of incorporation, most fillers are rubber-filled fragments. Still, as dispersion progresses, these agglomerates break down further, releasing the rubber trapped between particles into the matrix. As mixing progresses, dispersion reaches its limit, because rising temperatures reduce the rubber's viscosity, causing shear forces to diminish to minimal levels. A well-dispersed mixture generally has lower viscosity and shows greater die swell than a less uniform blend [6, 15, 18, 19].

When the carbon black is incorporated into rubber by mechanical mixing, a significant amount of rubber becomes insoluble, owing to the strong interactions induced during mixing.

Applying shear forces to carbon black aggregates generates free radicals, which may react with polymer chains. The filler, especially in the case of carbon black, may enter into weak interactions with the rubber.

Distribution: The distribution phase spreads the particles evenly throughout the polymer matrix, without altering their size or physical characteristics. This phase relies on the repeated shearing and folding of the mix, which helps separate filler aggregates and minimizes variations in local filler concentrations [15]. Using a stock blender attached to the two-roll mill or rolling the stock and re-introducing it endwise can enhance distribution and improve homogeneity.

The compounds are usually mixed in two or more steps in internal mixers:

1. The polymer, carbon black, other fillers, plasticizers, activators and anti-oxidants are added. Anti-oxidants can prevent the degradation of rubber, especially at higher processing temperatures.

2. The curatives, both sulfur and accelerators, are incorporated. In the first mixing stage, there is high heat generation, due to the higher shear forces involved.

A lower temperature is required for the second stage to avoid scorching due to the curatives. In an internal mixer, dispersion occurs mainly in the high-shear region of the mixer, which is the clearance between the rotor tip and the mixer wall. The width of the rotor tip is of great importance for the total shear developed.

During mixing on two-roll mills, it is generally preferable to incorporate fillers and ingredients, excluding vulcanizing chemicals, in an internal mixer. Subsequently, sulfur and accelerators are added and mixed on the mixing mill.

2.3.2 Influence of Method of Incorporation of Carbon Black Via Latex Stage Incorporation

In conventional mixing, the mastication stage involves the breakdown of polymer chains, making it easier for fillers and other ingredients to be incorporated into the rubber matrix. Peptizers are commonly added to facilitate the breakdown of the elastomer chains. The carbon black agglomerates break down into aggregates, disperse and finally distribute themselves homogeneously in the rubber matrix. Process oil is added to help filler dispersion. Coupling agents used with silica help enhance silica's reinforcing effect in rubber. Silanes are used for this purpose, as they chemically interact with rubber and filler in a reaction known as silanization. These coupling agents reduce filler aggregation and enhance rubber–filler interactions, improving tensile strength, reducing heat build-up and increasing tear strength. Effective silanization requires mixing at high temperatures, typically around 130 °C or within the range of 145–150 °C during the final mixing stage, as reported in [20]. One silane coupling agent suitable for enhanced silica rubber interaction is TESPT {bis[(3-(triethoxysilyl)propyl] tetrasulfide}. The optimal mixing temperature for this coupling agent for rubber–silica interaction is 135–150 °C [21].

Generally, a high amount of mixing energy is required in conventional mill mixing to develop shear forces sufficient for optimum rubber–filler interactions.

Latex-stage incorporation of carbon black and silica fillers is yields better filler wettability and dispersion, with lower energy consumption. It is well known that good filler dispersion and good rubber–filler interaction result in better reinforcement characteristics.

Good distribution is obtained when carbon black is introduced into the latex as a colloidal dispersion; however, effective interaction between carbon black and rubber requires additional shear forces generated during mixing in conventional equipment, such as a mixing mill. Higher shear forces are experienced by the filler–rubber matrix in the absence or in the presence of a minimum amount of plasticizer. This is why, in conventional mixing processes, the total amount of plasticizer is not initially added along with the filler.

From a technological perspective, the primary challenges are fundamental incompatibility and difficulty in dispersing highly aggregated or agglomerated fillers within the rubber matrix. Introducing carbon black and silica as dispersions during the latex stage should address some of these issues effectively.

The mixing of rubber with fillers such as carbon black and silica presents several challenges.

Aside from being highly energy intensive, it pollutes the surrounding area and the workplace and there are associated occupational health problems related to inhalation of fine dust.

There are very few systematic reports on latex-stage incorporation of filler by a simple process involving chemical destabilization and coagulation of latex with the expectation of better filler dispersion and better filler reinforcement.

Incorporating fillers such as carbon black and silica during the latex stage can improve wettability and dispersion while reducing energy consumption. It is well known that good flow characteristics can be effected by good filler dispersion. Good filler dispersion and good flow characteristics contribute to enhanced mechanical properties. However, the mixing of rubber with fillers such as carbon black and silica presents several other issues.

This process also improves flow behavior and mechanical properties by enhancing filler dispersion, which reduces filler–filler interactions and enhances filler–rubber interactions.

2.3.3 Production of Latex Carbon Black Masterbatch Based on a Modified Coagulation of Natural Rubber Latex – Factory Scale

The production of a masterbatch is based on a modified coagulation process. The production of latex– filler masterbatches involves the incorporation of specific fillers in specific dose as dispersions at the latex stage [22]. In the case of carbon black, the dispersion in water is 20% and for silica, it is 25%. These dispersions are prepared separately by ball milling for 24 hours. The dispersions are gradually added into fresh, natural rubber latex of known dry rubber content (DRC). The latex is stabilized with an anionic surfactant, which varies with the DRC and type of latex used – whether fresh, concentrated or a blend of both. This process is done under stirring to achieve the desired filler content. The stirring is continued for about 2–3 minutes and the masterbatch is coagulated by adding 2% formic acid under stirring until the rubber– filler masterbatch separates from the latex, leaving a clear serum.

The latex–filler masterbatch is allowed to remain in the coagulant for about one hour, then processed as thin sheets using a series of smooth rollers and washed free of acid. The sheeted coagulum is dried at 70 °C until properly dried and pressed into 25 kg bales using a hydraulic press, similar to crumb rubber processing.

Adding fatty acid soap to fresh latex alters the colloidal stabilization system, shifting it from protein-based stabilization to fatty-acid-soap-based stabilization. The filler dispersions are homogeneously dispersed in fatty-acid-soap-stabilized latex and coagulate quickly by adding formic acid. The coagulated rubber retains a good portion of the fatty acid added to latex before coagulation, which is evident from the raw rubber properties of coagulated rubber. The acid number and acetone extract of rubber from fatty acid soap-sensitized latex are higher than in conventionally coagulated rubber. Due to the presence of fatty acids, the Mooney viscosity and initial plasticity (P_0) are lower than those of conventionally coagulated rubber, as shown in Table 2.3. However, the Mooney viscosity can be improved by suitable chemical treatment during the processing of carbon black masterbatch. A schematic representation is shown in Figure 2.5 and factory images of the coagulation of the coagulation of fatty-acid-soap-sensitized latex mixed with carbon black dispersion, is shown in Figure 2.6.

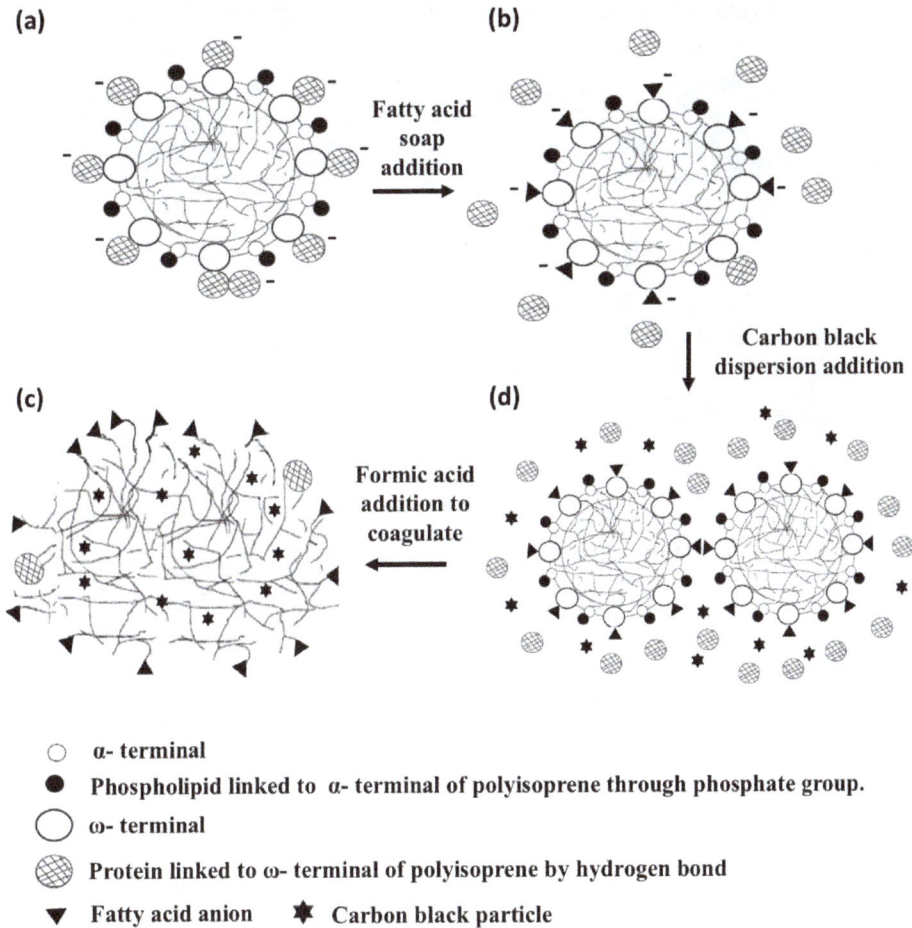

Figure 2.5 Proposed model of coagulation of latex in production of latex carbon black masterbatch: (a) protein-stabilized rubber particle, (b) fatty-acid-soap-stabilized rubber particle, (c) latex containing carbon black dispersion, (d) coagulated latex carbon black masterbatch

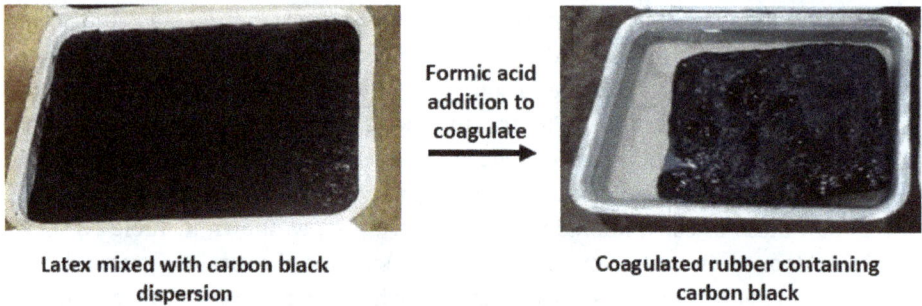

Formic acid
addition to
coagulate
→

Latex mixed with carbon black   Coagulated rubber containing
dispersion   carbon black

Figure 2.6 Factory images of the coagulation of latex carbon black masterbatch
(Courtesy: Ms Asiatic Rubbers, Kottayam, Kerala)

Table 2.3 Raw Rubber Properties of Rubber Obtained by Fatty Acid Soap Sensitized
Coagulation and Conventional Coagulation of Fresh, Natural Rubber Latex

Parameter	Rubber Obtained from Fatty-Acid-Soap-Sensitized Coagulation of Fresh Field Latex	Rubber Obtained from Conventional Coagulation of Fresh Field Latex
Acetone extract, %	4.5	1.9
Acid No. from acetone extract*	714	144
Mooney viscosity ML 1+4 at 100 °C	77	84
Initial plasticity	39	47
Plasticity retention index (PRI)	74	85

* The acid number is expressed as the number of milligrams of caustic potash required to neutralize the
acids present in the extract from 100 g of rubber. Acetone extract from about 10 g of rubber is
dissolved in 100 mL ethyl alcohol and titrated against 0.1 N KOH using phenolphthalein as an indicator.

A salient feature of latex masterbatch is that silica and different grades of carbon
black can be added in different concentrations [22]. Good filler dispersion also results
in good polymer filler interaction and better mechanical properties.

The filler dispersion for the latex masterbatch rubber compound (MBH 25/25, as per
the formulation in Table 2.4 and the dry rubber mixed rubber compound (Control
25/25, as per the formulation in Table 2.5 were determined with a Dispersion Analyzer from Tech Pro USA. The results are shown in Table 2.6 and Figure 2.7 .

Table 2.4 Formulation of the Latex Masterbatch Mbh 25/25 Containing Fillers HAF Carbon Black and Precipitated Silica

Ingredients	Dry Weight	Wet Weight
38% Fresh field latex	100	263
20% HAF carbon black dispersion	25	125
25% Precipitated silica dispersion	25	100
20% Fatty acid soap	As required	

Table 2.5 Formulation of Rubber Compound for Latex Masterbatch and Dry Rubber

Ingredients	Phr	
	MB25/25	Control
Natural rubber masterbatch	150	--
Dry rubber (ISNR5)	--	100
Stearic acid	1.5	2
ZnO	5	5
Antioxidant TMQ(2,2,4-Trimethyl-1, 2-dihydroquinoline, polymerized)	1	1
Antiozonant 6PPD(N-(1,3-dimethylbutyl)-N'-phenyl-p-phenylenediamine)	1	1
HAF carbon black	-	25
Precipitated silica	-	25
Naphthenic oil	-	5
N-cyclohexyl-2-benzothiazole sulfenamide (CBS)	1.0	1.0
Sulfur	2.5	2.5

Table 2.6 Filler Dispersion Rating

Sl No.	Sample Name	Filler Dispersion (X)[*]	Agglomerate Dispersion (Y)[*]
1	MBH 25/25	8.8	9.3
2	Control 25/25	7.0	9.2

* Dispersion is rated numerically from 1 to 10, with higher values indicating better filler dispersion.

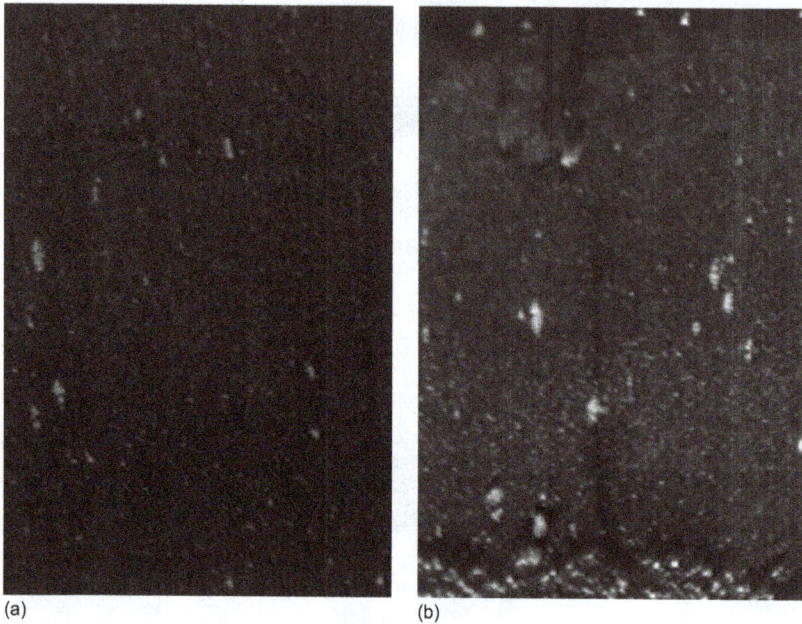

Figure 2.7 Filler dispersion images from vulcanizates using the Dispergrader:
a) MBH 25/25, b) Control 25/25

2.4 Shaping and Vulcanization

Generally, rubber is vulcanized by compression, transfer and injection molding. Compression molding is the most popular, because it is a very simple process that is accomplished with a hydraulic press.

The fillers added to rubber, their dispersion and interaction, such as filler–filler and rubber–filler, affect flow during shaping operations. The filler incorporation increases the viscous response. As a result, die swell generally decreases and the surface finish of calendered molded and extruded becomes smooth and affects shrinkage-related characteristics. Filler dispersion plays a major role in controlling the viscous and elastic responses during shaping operations, such as molding, extrusion and calendering. There are several instruments for evaluating the die swell and flow parameters of rubber compounds. These include rotational and oscillating-die viscometers and capillary rheometers.

2.4.1 Compression Molding

The efficiency of this process depends on the temperature, time and pressure employed. The heated mold cavity is filled with uncured rubber compounds. This uncured rubber compound takes on the shape of the mold cavity (about 5–10% of an extra compound may be used) to fill the mold (the compound weight is based on the volume of the mold). The rubber mold and the platens of the hydraulic press are closed. The mold is kept closed under pressure until the required vulcanization level occurs. The rubber compound is compressed under suitable pressure and temperature to ensure it fills the entire cavity. The pressure and heat cause the rubber to flow, taking on the mold's shape as it cures.

The force per unit area applied to the mold needs to be sufficient to allow the flow of rubber compound to fill the mold, to keep the mold pressed together so that flash can flow out through the flash channel and to take care of the gases (to prevent opening of the mold) released during the production of expanded rubber. In typical rubber compounds, molding pressure can be 120–140 kg/cm^2. The shear rates during flow are 1–10 s^{-1}.

2.4.1.1 Mold Shrinkage During Compression Molding

A problem with compression molding is that there will be a change in the product's dimensions and shape, due to shrinkage after removal from the mold. Shrinkage describes the difference in dimensions of the mold and the articles produced from it when both are measured at room temperature. To an extent depending on the polymer type, linear shrinkage generally ranges from 1.5–3%. The coefficient of thermal expansion of mold materials such as steel or aluminum is lower than that of the rubber compound contained in the mold. So, during molding at high temperature, expansion of the rubber is limited by the mold and, during cooling, contraction happens such that the size will be less than the corresponding mold cavity. As the rubber expands during heating at high temperatures, the pressure in the mold rises, leading to backrinding in the product (tearing at the mold partition line) and fracturing of weak molds. In constrained conditions, rubber is vulcanized, but, on removal and cooling, the rubber undergoes a higher degree of linear contraction than the mold, and shrinks.

The proportion of fillers within a rubber compound also has a significant effect, principally because their coefficient of linear expansion is closer to that of steel than rubber. Additives such as fillers and recycled rubber reduce shrinkage.

2.4.2 Extrusion

Extrusion is a process by means of which strips of material are shaped to a desired cross-section and any length by forcing the material through an orifice or die under controlled temperature and pressure conditions. Screw extruders are used for extrusion of rubber. To reduce premature cross-linking, the L/D ratio of extruder barrels is less than that of thermoplastics, and is typically 10–15. In an extruder, the rubber compound must move through the barrel and be pushed into the die head under pressure so that the extruded material has a smooth surface. Movement of the rubber compound is accomplished in a single-screw extruder mainly by the frictional forces developed in the barrel.

Extrusion process: Rubber compound is pre-warmed and forced under pressure through an orifice (die) that shapes the extrudate to a desired cross-section, as shown in Figure 2.8. Die swell occurs, due to elasticity. Polymers are long macromolecules with entanglements. Consequently, they offer high resistance to shearing forces.

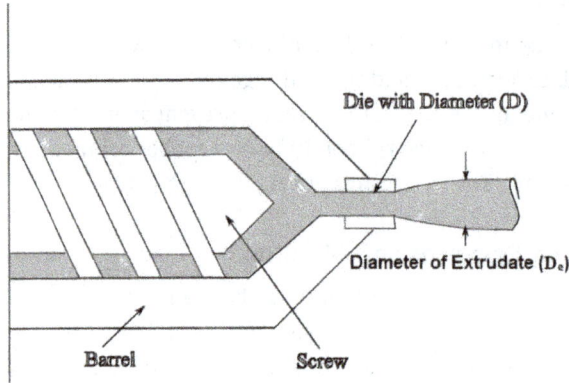

Figure 2.8 Schematic representation of the extrusion process

The rubber compound of suitable viscosity is fed into the feed hopper of the extruder. The screw begins to carry the rubber forward into the die and, due to this movement, the pressure exerted on the rubber increases. The viscosity of the rubber reduces when it moves forward and enters the die head, because the die head is maintained at a higher temperature, which ensures a good flow of rubber as the rubber is being pushed through it.

The shear rate experienced by the rubber compound is high while it is flowing in the extruder. If filler is added, more plasticizers can be added to maintain the required viscosity during processing. The compounding requirements are such that elastic recovery needs to be reduced.

Elastic recovery is reduced by increasing filler loading or by adding factice, a processing aid. Partially cross-linked, superior processing rubber can also be incorporated into the formulation to reduce elastic recovery. Extrudability is improved with reinforcing fillers and heat generation is also increased during mixing. Process aids assist in promoting a smooth surface. Waxes, fatty acids or fatty acid derivatives are useful additives. Too soft a compound can lead to a loss of extrusion efficiency. The quantity of plasticizers used varies with the rubber product.

The die is not at a very high temperature during extrusion; this avoids cross-linking of the rubber. The uncured rubber compound is extruded and then vulcanized.

During extrusion, the chain molecules are aligned inside the extrusion die to a new, suitable geometry, but the rubber molecules do not have sufficient time to adjust to this new geometry. Upon emerging from the die, the strained rubber molecules partially recover to the original unstrained condition. This elastic recovery is responsible for the greater diameter of the extruded rubber compared with that of the die. The ratio of extrudate diameter to die diameter is called the die swell. Both surface finish and die swell are critical in rubber processing. Extrudate shrinkage is another way of expressing die swell and the nature of these two parameters depends on the visco-elasticity of rubber.

When the rubber is in the die, during the deformation process, some of the energy required for deformation is stored by virtue of the elastic component of rubber. This stored energy is released when the rubber exits the die, leading to some recovery on the part of the elastic component. Simultaneously, time-dependent deformation or flow occurs, due to the viscous component, represented as a dashpot.

In the rubber industry, shrinkage (die swell) is a serious concern, as many products are made using extruded profiles that are later assembled before vulcanization. Examples are tires, hoses and beltings.

Polymers, in addition to extrudate swell, exhibit a property called rod climbing or the Weissenberg effect. The cause of the Weissenberg effect, like that of die swell, is attributed to the elasticity of rubber (in the liquid state or solid state).

2.4.3 Calendering

Two or more rolls or bowls are assembled on a framework in a conventional rubber calender, producing sheets of various thicknesses, coating textiles or fractioning fabrics with rubber compounds. A photograph of an industrial three-roll calender is shown in Figure 2.9. As already mentioned, the gap between the rolls is called the nip gap. The compound viscosity is adjusted to yield smooth sheets or coatings by heating and cooling the compound. The calender rolls are made of appropriate metals, such as solid cast iron, to combat the shear stress developed during calendering. Usually, in a three-roll calender, the central roll is driven and the upper and lower rolls are driven by gears with the help of the central roll. The two rolls may rotate at the same speed for sheets or topping but, for frictioning, the rolls operate at uneven speeds.

During calendering, a rolling bank forms at the nip and the mechanical forces acting on the rubber can lead to tearing and entrapped air. The material is subjected to high compression force to remove the air, which is why calendered sheets are generally thin. In calendering, the unvulcanized rubber compound passes through a series of progressively smaller gaps between sets of rolls rotating in opposite directions. The final nip gap determines the thickness of the calendered sheet.

The compound's viscosity is adjusted for the purpose of achieving a smooth coating or sheeting process; the compound passes through a warming mill to ensure proper viscosity is maintained before it is loaded onto the calender. The shear rates are higher than in compression molding, but may be lower than in extrusion. A higher amount of plasticizer is added to ensure higher flow during the processing of uncured rubber compound to thin uniform sheets. In frictioning operations, the stock must be soft and tacky to penetrate the fabric's interstices easily. NR and CR compounded with resins such as coumarone indene or petroleum types yield good frictioning stocks.

Viscoelasticity is important in the calendering process. Good flow (viscous component) can result in smooth sheets but, if the viscous component predominates in the rubber, it may become tacky on the rolls. If the elastic component is high, the shrinkage rate of calendered sheets will be high, and so the film's smoothness will decrease.

Figure 2.9 Photograph of a three-roll calender (Courtesy: Glenrock Rubber Products Pvt. Ltd, Kerala, India)

2.5 Vulcanization of Extruded and Calendered Profiles

The vulcanization methods employed for extruded and calendered products broadly fall into the categories of batch and continuous techniques. The use of autoclaves or lead-curing, batch vulcanization offers precise dimensional control, but is time-consuming and often less efficient for large-scale production. On the other hand, continuous vulcanization techniques have revolutionized the industry by offering faster, more uniform curing and greater productivity. Devices such as hot air tunnels, fluidized beds, liquid-curing methods, continuous drum cures and microwave vulcanization have enabled manufacturers to meet diverse application requirements with improved cost-effectiveness and reduced energy consumption. These techniques are tailored to specific products, balancing speed, uniformity and surface quality while addressing challenges such as porosity and material compatibility. The choice of vulcanization method plays a crucial role in determining the performance and application of the final rubber products, highlighting its importance in modern rubber processing.

2.5.1 Batch Vulcanization Techniques

2.5.1.1 Autoclave or Steel Pan

This is a cylindrical pressure vessel and may be jacketed or unjacketed. Usually, the components are given some support to prevent shape distortion during curing.

2.5.1.2 Lead Curing

The extruded profile is given a lead covering by means of a lead extruder. The inner portion is filled with water and sealed at the ends of the hose. When heated in an autoclave, the water squeezes the extrudate against the walls of the lead sheath. The lead sheath is stripped away after vulcanization and cooling clamps are removed.

2.5.2 Continuous Vulcanization Techniques

Continuous vulcanization techniques have revolutionized the process of vulcanizing extruded and calendered products. By using continuous vulcanization techniques, manufacturers can increase production efficiency and reduce the overall costs associated with vulcanization. Conventional techniques used for batch vulcanization include autoclaves, hot air ovens and water curing. However, these have their limitations and can be time-consuming and expensive.

2.5.2.1 Hot-Air Tunnel

This method uses a conveyor belt to carry rubber extrudates through a heated-air oven for continuous curing. The air inside the oven can be heated by means of either electrical or gas heating methods. While cost-effective for small, slow-running profiles, hot air suffers from poor heat-transfer efficiency and requires longer production lines. It may cause oxidation of the surface during curing. Despite its disadvantages, the hot-air tunnel is still viable for vulcanizing extruded and calendered products with low investment needs. In addition, it yields a clean and attractive surface. Hot-air tunnels are most suitable for cellular products and carpet underlays where close dimensional tolerances are not very critical.

2.5.2.2 Fluidized Bed

This technique employs a preheated gas that keeps tiny glass beads in constant motion. The high velocity of the steam/compressed air fluidizes the glass beads, creating a bed into which the rubber profiles submerge. The outcome is faster cure cycles, due to improved heat transfer compared with a hot-air tunnel. However, the beads can stick to the rubber and this limits applications for delicate profiles.

2.5.2.3 Liquid-Curing Method (LCM)

An insulated tank filled with a heated liquid medium (eutectic salt, glycol, oil or low-melting metal alloy) transfers heat to the rubber profiles. LCM enables rapid heat transfer, due to the high heat-transfer coefficient of the liquid, facilitating fast line speeds and short cure cycles. Eutectic salts are commonly employed on account of their high thermal conductivity, low viscosity and cost-effectiveness. Vulcanized products have to undergo to salt removal, washing and drying. Careful control and speed synchronization are essential for achieving precise dimensional control.

2.5.2.4 Continuous Drum Cure

This method is commonly employed for the continuous vulcanization of calendered rubber sheets or fabric-reinforced products, such as conveyor belts and rubber roofing/flooring materials. It involves passing the rubber material to a heated drum with the help of a tension belt, which presses the material against the drum during vulcanization (Figure 2.10). By virtue of the direct contact between the rubber material and the heated drum surface, the continuous drum cure method offers excellent heat transfer and uniform curing. Sheets with embossed designs and patterns can also be produced by using customized drums with specific patterns, as shown in Figure 2.11.

Figure 2.10 Embossed sheets of different designs (Courtesy: Glenrock Rubber Products Pvt. Ltd, Kerala, India)

Figure 2.11 A typical industrial continuous drum-curing machine (Rotocure) (Courtesy: Glenrock rubber products Pvt. Ltd, Kerala, India)

2.5.2.5 Microwave Vulcanization

The microwave vulcanization method is a relatively new technique that uses electromagnetic radiation in the microwave frequency range to generate heat and cure rubber profiles. It is ideal for polar rubber compounds or those rendered polar through the incorporation of specific fillers and additives. Microwaves can penetrate through the material, interact with the polar molecules and generate heat directly within the rubber, resulting in faster and more uniform curing than traditional methods.

Microwaves utilize the principle of dielectric heating (Figure 2.12). When a polar material is subjected to an alternating field, the molecules align themselves to the field. When the field is reversed, the alignment of the molecule also reverses. At higher frequencies, this happens at a higher rate and the molecules often fail to stay in phase with the field; consequently, part of the energy is dissipated as heat.

Polar rubbers, such as NBR, CR, CSM, fluorinated rubbers, acrylic rubbers, etc. can generate heat when exposed to microwaves. Non-polar rubbers, such as EPDM, SBR, NR, etc., must be modified with any of the additives PEG, triethanolamine, carbon black, silica, etc. to increase their ability to generate heat when exposed to microwaves. Ingredients such as zinc oxide, stearic acid and magnesium oxide can also be added to enhance the heat-generating properties of the rubber [23].

Microwave vulcanizers are generally combined with hot air in continuous vulcanization. Microwaves generate the heat necessary for vulcanization, while the hot-air tunnel sustains this heat, preventing the profile from cooling. This approach makes for a cost-effective and efficient vulcanization process, optimizing resource utilization and productivity.

Figure 2.12 A prototype microwave vulcanizer (Courtesy: NIRT, Rubber Board, Kerala)

Automotive profiles, such as weather strips and seals, sponge extrusions for HVAC systems and profiles for building and construction utilize microwave vulcanization, because it is faster and offers more uniform curing.

Given that all these methods are pressureless, there is a significant risk of product porosity. During pressureless heating, volatile materials such as moisture in the compounding ingredients may vaporize during vulcanization, but remain trapped in the rubber matrix, occupying space and giving rise to porosity in the final product. It is advisable to use calcium oxide as a desiccant during the compounding process to avoid the problem of porosity. Typically, the desiccant is dispersed in oil and enclosed in air-tight covers to prevent premature reaction with moisture-laden air.

In summary, continuous vulcanization offers notable advantages over traditional batch methods by enhancing production efficiency and lowering costs. Several techniques are available, each with its own set of strengths and weaknesses. Hot-air tunnels are less expensive, but curing takes longer, while fluidized beds cure more quickly, but carry the risk of surface irregularities. Liquid curing allows for rapid heat transfer, but requires post-processing. Continuous drum cure ensures uniform curing for sheet products and microwave vulcanization provides fast and uniform vulcanization of suitably formulated compounds.

2.6 Viscoelastic Parameters of Raw Rubber and Characterization

While synthetic rubbers are engineered to achieve a desired molecular weight, microstructure and macrostructure, natural rubber has a molecular structure that is formed during biosynthesis. It has unique properties that stem from its specific micro- and macrostructure.

2.6.1 Microstructure of NR

The rubber contains non-rubber ingredients, such as lipids, proteins and other ingredients retained on the rubber during the conversion of liquid latex to solid rubber. Table 2.7 shows the composition of raw natural rubber sheet. These non-rubber ingredients play a specific role in the quality of vulcanized rubber.

Chemically, this rubber molecule is cis-1,4-polyisoprene (contains more than 99.9% of cis-1,4- structural units) and is completely stereoregular. It has a very high molecular weight, whereby the exact weight depends on the clones and the age of rubber trees, as well as the soil and climate conditions of the plantation. Natural rubber molecules have linked fatty acids originating from the phospholipids. These phospholipids and these fatty acids are known to help with strain-induced crystallization, a significant contributing factor to the outstanding gum strength of vulcanized natural rubber.

Table 2.7 Composition of Natural Rubber Ribbed Smoked Sheet

Component	%
Moisture content	0.6
Acetone extract	2.9
Protein calculated from nitrogen content	2.6
Ash content	0.4
Rubber hydrocarbon	93.5

2.6.2 Strain-Induced Crystallization of Natural Rubber

Rubber molecules tend to align themselves to form a level of orderly arrangement and they form crystalline structures when cooled to a temperature close to -30 °C. This is called the temperature-induced crystallization (TIC) of natural rubber. This crystallization vanishes after vulcanization and is of minor significance in rubber processing and product manufacture, as most rubber products are used after vulcanization. Several reports show that the temperature-induced crystallization of natural rubber is affected by the non-rubber ingredients present in the rubber. Rubber molecules appear linked to phospholipids containing long-chain fatty acids in rubber particles of NR latex. Thus, raw rubber contains long-chain fatty acids covalently linked to the chain end. Saturated fatty acids induce crystallization of the rubber chain, while unsaturated fatty acids – present in NR as a mixture – act as plasticizers and act synergistically with saturated fatty acids to accelerate crystallization of rubber [24].

The rubber molecules have a level of orderly arrangement even after vulcanization when rubber is in a strained condition and this crystallization is called strain-induced crystallization (SIC). This type of crystallization happens when rubber is stretched beyond a certain point and, as a result, the strength of the rubber increases. SIC is observed as the abrupt increase in stress when strain is applied to natural rubber to an extension of about 200%. At very high strain rates, the stress at a given strain increases, and the onset of SIC shifts to a lower strain. SIC limits movement between neighboring chains and results in a type of reinforcement that leads to enhanced modulus and hysteresis. On stretching, the rubber chains appear to be locked in certain positions. On release, the crystallization disappears.

SIC is attributed to the very high molecular weight of and stereoregularity in the long chains of cis-polyisoprene. However, some reports show that entanglements and pseudo-linked networks formed by the interaction of functional groups at the chain end with non-rubber components can play a role in the strain-induced crystallization of natural rubber [25, 26]. With increase in temperature, SIC decreases, with the reduction being very notable at about 80 °C [27].

One of the main reasons for the excellent mechanical strength (even in the absence of reinforcing fillers) and natural rubber's crack growth resistance is its ability to undergo strain-induced crystallization. The growth of a crack is arrested, because crystallization occurs at the tip of the crack where strains are high. Even in dynamic conditions, the crack-growth resistance of NR is high, owing to strain-induced crystallization. In particular, it is believed that the induced crystallites retard, deviate and even stop crack growth under static or cyclic loading conditions [27, 28]. This property also results in good wear resistance. Consequently, natural rubber has many specific applications in static and dynamic conditions that require high mechanical strength coupled with hysteresis. This explains its use in products intended for engineering applications, such as seals, gaskets, tubes and hoses, automobile tires and vibration and shock isolators.

Synthetic polyisoprene (IR) is primarily cis-1,4-polyisoprene, but contains small amounts of 3,4- and trans-configurations. This rubber does not show strain-induced crystallization to the extent shown by natural rubber. It is accepted that IR and natural rubber have more or less similar properties, but differences in stereoregularity. For practical purposes, natural rubber is considered to be almost 100% cis-polyisoprene, whereas IR has 95–98% cis-polyisoprene, along with minor amounts of other isoprene configurations. However, this slight difference in stereoregularity affects the mechanical properties and strain-induced crystallization. IR is produced by polymerization reactions involving titanium (Ti) and lithium (Li) as catalysts. When Ti serves as the catalyst, the trans-configuration is 1% and the 3,4-configuration is 0.5%.

In contrast, with Li acting as catalyst, the trans-configuration increases to 5%, while the 3,4-configuration reaches 5%. Unlike natural rubber (NR), synthetic polyisoprene does not contain non-rubber ingredients, such as proteins, carbohydrates or phospholipids. As a result of these factors, the strain-induced crystallization behavior of IR is generally less effective than that of NR [29].

2.6.3 Macrostructure

Macrostructure refers to the molecular weight, branching and gel formation of rubber. Molecular weight can be expressed as the number-average molecular weight (M_n), weight-average molecular weight (M_w) and molecular weight distribution (MWD). NR exhibits a very high molecular weight and is known generally for its bimodal molecular weight distribution.

The gel is defined as a network of rubber chains which is formed through extensive branching or cross-linking and is usually swollen by solvents, but does not dissolve in them. Two types of gel exist in NR: macro gel and micro gel. Macro gel (macro aggregates) is that part of NR which is visible and insoluble in a conventional polyisoprene solvent and can be eliminated by centrifugation. Microgel (micro aggregates), contained in the soluble part, cannot be eliminated by centrifugation.

The presence of gel in rubber coagulated from fresh latex prevents it from being fully soluble in toluene. Notably, latex obtained from trees tapped for the first time exhibits a significantly higher gel content than that from trees tapped regularly. The gel remains in the latex even after preservation and storage. Thus, natural rubber can be considered to consist of two components: a solvent-soluble phase (sol phase) and a solvent-insoluble phase (gel phase).

Studies conducted on the gel phase have revealed a higher concentration of nitrogen-containing compounds than the sol phase; this suggests that rubber chains are interlinked through interactions involving proteins and phospholipids present in adjacent rubber molecules [30, 31]. Gel formation in latex is attributed to the network structure formed by rubber chains [32] interconnected through these non-rubber components. These components – primarily proteins and phospholipids – interact at the α- and ω-terminals *via* hydrogen bonding or ionic bonds involving metal ions [33]. The functional groups at these terminals initiate branching and entanglements, facilitating cross-linking and the development of a network that leads to gel formation, as shown in Figure 2.13.

Branching in NR molecules is believed to consist in hydrogen bonding between polar groups of phospholipids. Branching is also possible through the formation of cross-links by cationic linkages between the negative charge of phospholipid and divalent cations (Mg^{++} ions). The specific end groups help in chain entanglement.

The unique macrostructure of natural rubber, characterized by the ability of rubber molecules to have long-chain branching, specific interactions between polar groups attached to rubber chains, chain entanglements and the presence of gel, significantly contributes to its high green strength. The factors contributing to macrostructure also affect the flow characteristics of NR [34].

Increase in viscosity during storage

Natural rubber shows storage properties that are distinct from those of synthetic polyisoprenes. Natural rubber tends to harden gradually over a few months of storage, a phenomenon known as storage hardening, while synthetic rubber typically softens. This is of considerable importance, since it affects the processing properties. It is known that there are certain abnormal groups on rubber chains, such as epoxy, aldehyde, carboxyl, etc. on the rubber molecules [29]. Hardening is believed to be due to cross-linking reactions involving abnormal groups, such as carbonyl and epoxide in the rubber molecules, and non-rubber ingredients, such as proteins, especially amino acids, in the rubber. This slow cross-linking is accelerated by anhydrous conditions [35]. Storage hardening affects processing properties such as Mooney viscosity and Wallace plasticity [36, 37].

a)

b)

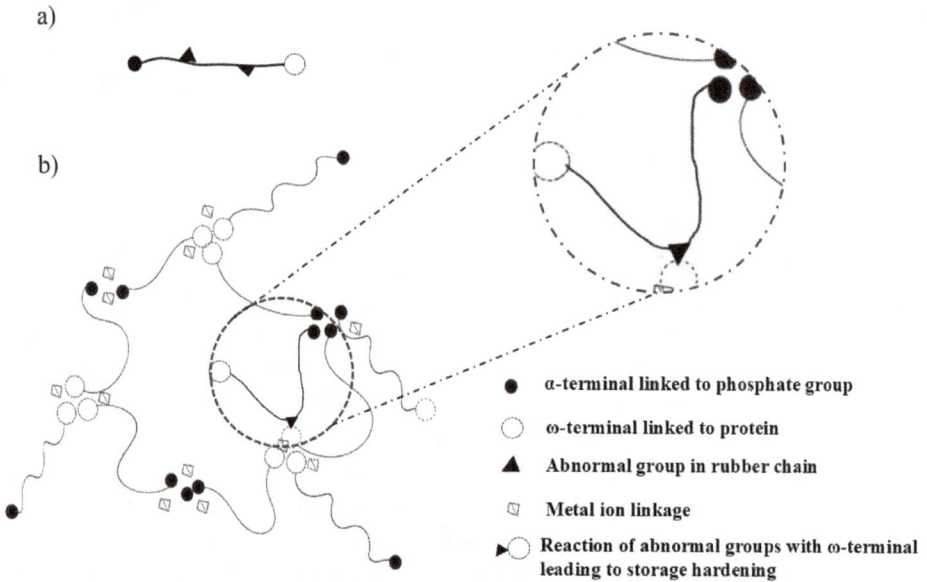

- ● α-terminal linked to phosphate group
- ○ ω-terminal linked to protein
- ▲ Abnormal group in rubber chain
- ⊡ Metal ion linkage
- ▶○ Reaction of abnormal groups with ω-terminal leading to storage hardening

Figure 2.13 Schematic representation of (a) a rubber molecule with an abnormal group, (b) branching leading to gel formation and storage hardening due to abnormal groups in natural rubber

2.7 Viscoelasticity of Raw Rubber and Rubber Compounds

Rubber rheology can provide insight into viscoelastic parameters, which are heavily influenced by molecular chain motion and the degree of entanglement. The latter can be visualized as polymer chains wrapping around each other. As entanglement increases, the chains exhibit greater resistance to deformation and flow, since more energy is required to disentangle the chains and enable their relative movement. This resistance is closely tied to molecular weight: above a threshold known as the critical molecular weight, viscosity rises sharply with increase in chain length.

Viscoelastic behavior is sensitive to strain rate and temperature. At higher strain rates, the flow is not smooth. The effect of temperature is that, as the temperature increases, the kinetic energy of the rubber molecules increases, leading to an increased ability to flow. This increase in temperature also enhances the ability of polymer chains to slip past one another.

It is well known that the lower the glass transition temperature (T_g) of rubber, the more rubbery it becomes at room temperature. The viscosity of rubber and rubber compounds is sensitive to the difference between its processing temperature and its T_g. The greater the difference, the less the effect of temperature on its viscosity will be.

2.8 Instruments Used to Measure the Viscoelasticity of Natural Rubber in Relation to Processing

The rheological behavior of rubber involves both Hooke's law of elasticity and Newton's law of viscosity. The factors governing the viscoelastic behavior of rubber during flow depend on the nature of the rubber and rubber compound and factors related to flow and temperature. The microstructure and macrostructure of rubber, the presence of different additives, such as curatives, flow characteristics, such as shear stress shear rate, duration of application, deformation force and temperature affect the flow of rubber. The flow aspects are measured in the form of different parameters reflecting viscous and elastic components, such as plasticity, viscosity, die swell, elastic modulus, viscous modulus and loss tangent. This is done with various types of instrument, such as compression polarimeters, rotational viscometers, capillary rheometers and dynamic mechanical analyzers.

2.8.1 Compression Plastimeter: Wallace Rapid Plastimeter

Plasticity measurements are based on rubber samples' response to radial flow and molecular relaxations under a compressive force. The Wallace rapid plastimeter measures the viscoelastic characteristics of raw rubber, such as initial plasticity (P_0) and plasticity retention index (PRI), which indicates the resistance of rubber to oxidative degradation. PRI helps with our understanding of the processing behavior of rubber and the product properties after vulcanization.

P_0 of rubber is sensitive to various factors, such as clone, micro- and macrostructure and storage of rubber, in addition to the chemicals used in primary processing of NR. Rubber with high molecular weight can have a high P_0.

The initial plasticity can be enhanced by treating the wet coagulum with chemicals such as hydrazine hydrate (HH) and tetraethylenepentamine (TEP); the rubber interacts with these chemicals to form a gel. Both P_0 and PRI are improved by treating raw rubber coagulum with tetraethylenepentamine.

The basic principle of measurement is that the rubber flows under the application of a constant force and the flow can be measured from the change in thickness of the sample before and after the application of force.

The raw rubber sample is homogenized by passing it six times through the nip of a two-roll mill (maintained at a suitable nip gap). Then, it is passed through a tight nip and folded to become a sheet of raw rubber. A specimen cutter is used to cut out test samples as pellets so that the P_0 and PRI may be determined.

Accelerated aging is conducted by conditioning the pellets in an aging chamber maintained at a temperature of 140 °C for 30 minutes. A sample is compressed between two circular platens and maintained at a temperature of 100 °C. The sample is conditioned for 15 seconds at a thickness of 1 mm. A compressive force of 100 N is then applied for 15 seconds. The final thickness of the test piece, expressed in units of 0.01 mm, is the plasticity number. Plasticity tests are also done on aged samples; usually, three test pieces are tested. PRI is calculated as follows:

$$\text{PRI} = (P_{30}/P_0) \times 100 \qquad\qquad (2.10)$$

Where

P_{30} = Aged plasticity

P_0 = Initial plasticity

This test serves as a quality control parameter for raw natural rubber.

2.8.2 Rotational Viscometer: Mooney Viscometer

The Mooney viscometer measures viscosity in Mooney units by recording the torque required to rotate a disk-shaped rotor inside a chamber filled with rubber. The applied shear rate is low, typically 1 s^{-1} (Appendix 1). Mooney viscosity is recorded by measuring the combined effect of shear rate, temperature and shear stress as a function of time.

The Mooney viscometer is a shearing disk viscometer designed to accurately measure elastomers' viscosity, scorch time and cure characteristics. The instrument consists of a flat, cylindrical disk driven by a motor to rotate slowly and continuously in one direction at two revolutions per minute. There are two types of rotors: large and small. The large rotor is preferred unless the Mooney viscosity exceeds the instrument's torque capacity or if slippage occurs or is suspected [38]. This disk is embedded into an elastomer specimen confined in a heated die cavity maintained at a specified temperature and closed under a specified force. As the disk rotates, it experiences a shear strain. This resistance to rotation offered by the elastomer is shearing viscosity, and is proportional to the mean absolute viscosity of the specimen. The disk is grooved to avoid rubber slippage. Normally, in this test, the standard temperature as per ISO 289 is 100 °C.

The viscosity is expressed as ML (1+4) at 100 °C. The Mooney viscometer is also used to measure scorch characteristics (ASTM D 1646). For this, the temperature is set to 120 °C. Due to the temperature, the viscosity initially decreases, passes through a minimum and starts increasing. The time required to increase Mooney viscosity by 5 units is taken as the scorch time at 120 °C and the time required to obtain a rise of 30 units is taken as the cure time at 120 °C.

Storage hardening can cause a change in molecular weight, an increase in Mooney viscosity and P_0 and a decrease in PRI. Samples with higher Mooney viscosity (ML (1+4) at 100 °C) require a longer time for mastication and are sometimes peptized to lower the viscosity of NR to the level required for subsequent mixing operations.

2.8.3 Oscillating Rheometers, Moving-Die Rheometers and RPA 2000

Common instruments for determining cure aspects are the oscillating disk rheometer (ODR) and the moving-die rheometer (MDR), which are technically classified as cure meters. The instruments yield a plot of torque versus time as a measure of cure characteristics.

Rubber is placed in a cavity formed by two circular dies. One of the dies oscillates, exerting a dynamic strain, and the torque transmitted is measured on the spindle of the other die. The phase difference between a dynamic strain and the response is called the phase angle δ. The viscoelastic parameters are evaluated by varying the angular shear amplitude (±0.05 to 90°) or a shear (±0.07% to 1256%) or the frequency (0.1 to 2000 cycles/minute) and temperature. Figure 2.14 shows a typical rheograph generated with a moving-die rheometer.

The instrument measures complex torque, which arises from the viscoelastic nature of rubber and is not in phase with the applied strain. This torque comprises an elastic component, which is in phase with the strain, and a viscous component, which 90° out of phase. The dynamic viscosity measured by RPA can be correlated with viscosity values measured with a capillary rheometer. The instrument also measures processability parameters, such as minimum viscosity, maximum viscosity, scorch time, cure time, cure rate and reversion time (for natural rubber).

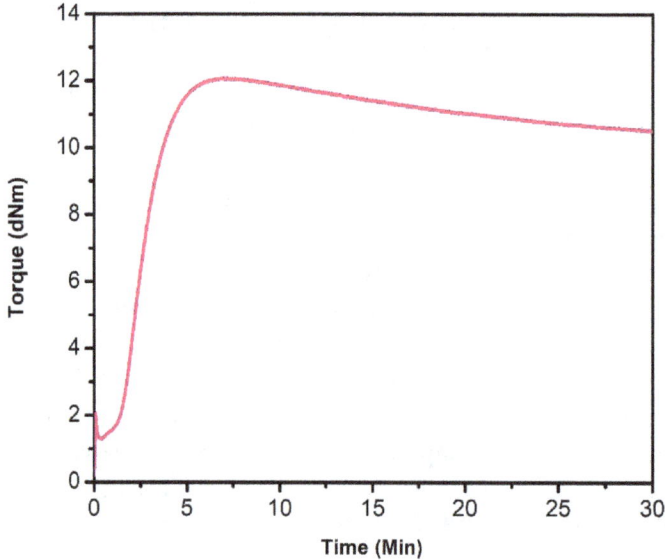

Figure 2.14 Typical rheograph obtained from a moving-die rheometer

2.9 Measurement of Processability as Related to Extrusion

A capillary rheometer returns information on steady shear flow at shear rates corresponding to different processing operations used in rubber processing. The rubber capillary rheometer determines the optimum working parameters of rubber materials for the purpose of assessing flow in compression molding, calendering and processing characteristics during extrusion. It can be attached to an Instron universal testing machine and is able to determine elasticity parameters, such as die swell, principal normal stress difference, recoverable shear strain and elastic shear modulus.

2.9.1 Capillary Rheometer

In capillary rheometry, the rubber compound is placed in a temperature-controlled reservoir and is forced through a small-diameter capillary tube. A plunger (piston) either applies a constant pressure (constant stress) or moves at a fixed speed (constant shear). The measurements are carried out at different piston speeds and the viscosity is obtained from the measured pressure drop ΔP and the measured volumetric flow rate Q, which is determined by the piston speed. The rubber is pushed through the capillary at a known flow rate. Rubber is filled in a barrel maintained at a specific temperature, compacted with a piston and extruded through a capillary die at different shear rates. The viscoelastic nature can be evaluated by observing the extrudates.

Since flow through a capillary is non-Newtonian, some correction is needed to obtain reliable data. The flow is not laminar, but plug flow, and so the shear rate obtained at the wall is an apparent value. The true shear rate at the wall is determined by applying the Rabinowitch correction. The pressure also drops during flow in the capillary, for which the Bagley correction is applied [39]. A schematic drawing of a capillary rheometer is provided in Figure 2.15.

Viscosity is affected by the micro- and macrostructure of rubber and, in the case of filled rubber, by the various rubber–filler and filler–filler interactions.

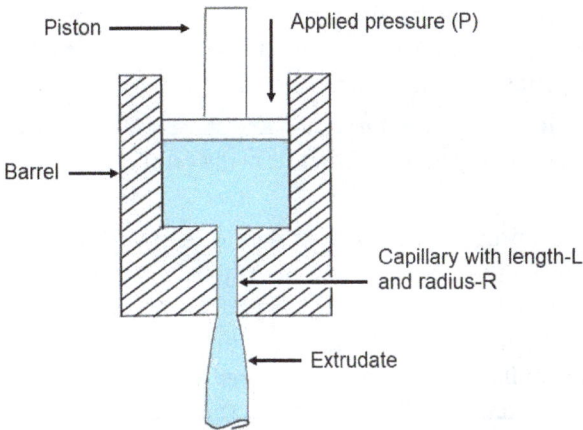

Figure 2.15 Schematic drawing of a capillary rheometer

During flow through a capillary, the shear rate varies with the distance from the capillary axis, so a point such as the capillary wall is selected for the determination of the shear rate and shear stress. As the rubber moves through the capillary, it experiences a pressure drop (ΔP), which increases linearly with length and corresponds directly to the wall shear stress, as shown below.

$$\tau_{wapp} = R\Delta P/2L \tag{2.11}$$

$$\dot{\gamma}_{wapp} = 4Q/\pi R^3 \tag{2.12}$$

$\dot{\gamma}_{wapp}$ is the apparent wall shear rate, τ_{wapp} is the apparent wall shear stress and R and L are the radius and length of the capillary, respectively.

Q is the flow rate of the rubber pushed through the capillary.

$$\text{Apparent viscosity} = \tau_{wapp}/\dot{\gamma}_{wapp} \tag{2.13}$$

For Newtonian fluids,

$$\dot{\gamma}_{wapp} = \dot{\gamma} \cdot w \tag{2.14}$$

For Non-Newtonian fluids, this relation does not hold and the shear rates at the wall may be higher.

Based on the Rabinowitsch correction, the true shear rate $\dot{\gamma}_{tw}$ is:

$$\dot{\gamma}_{tw} = [3\,n + 1/4n]\,\dot{\gamma}_{wapp} \tag{2.15}$$

Where

n = slope of the log-log τ_{wapp} versus $\dot{\gamma}_{wapp}$ flow curves and called the power-law index

$$n = \frac{\Delta \log R\Delta\,P/2L}{\Delta\log 4Q/\pi\,R^3} \tag{2.16}$$

This power-law index indicates how rapidly the viscosity decreases with increase in shear rate. For polymer fluids, n varies from 0–1. When $n = 1$, the flow is Newtonian and the consistency index corresponds to Newtonian viscosity.

When the polymer enters the die, it undergoes deformation and stores energy, due to its elastic component. It releases the energy as it flows out from the capillary, and this manifests as die swell.

When calculating the pressure drop in the capillary, it should be noted that the drop in pressure measured between the reservoir and the capillary outlet are affected by factors such as pressure loss upon entering the capillary and pressure loss upon exiting it. These must be included in the pressure drop experienced by the rubber flowing through the capillary. This flow is assessed with the aid of at least three capillaries of same diameter, but different length. The measured values of ΔP are plotted against L/R to yield a linear graph. The ΔP at $L = 0$ is called the Bagley correction (Pc) for obtaining true shear stress τ.

(Bagley's method treats the extra pressure drops at the ends as if the capillary were extended by a length $e\,L/R \rightarrow L/R + e$.)

$$\tau_{tw} = R(\Delta P - Pc)/2\,L \tag{2.17}$$

$$\text{True viscosity} = \dot{\gamma}_{tw}/\tau_{tw} \tag{2.18}$$

There is a term called recoverable shear strain γr that can be related to the elastic parameter as follows:

$$\gamma r = \sqrt{[2(De/D)^6 - 2]} \tag{2.19}$$

Where

De = Diameter of extrudate

D = Diameter of die

Die swell De/D increases with increase in shear rate and decreases under a fixed shear rate with increase in temperature. Die swell decreases with increase in land length, which is the l/d ratio of the capillary (where l is the length of the capillary and d is the diameter). Die swell decreases with increase in filler concentration.

2.9.2 Extrudability Performance

This can also be determined using a screw-type extruder equipped with an ASTM Extrusion Garvey Die head. The rubber material is extruded through a specially shaped die, producing an extrudate with relatively flat surfaces, sharp corners and thin sections. For the test, the extruder is pre-set to a suitable barrel die head and die temperature. The temperature settings and test conditions are based on ASTM D2230–96:2002. To obtain extrudability performance, carbon black and silica-filled rubber compounds can be assessed using die swell-related parameters.

2.9.3 Extrudability Performance of Rubber Mixes

The samples contain different loadings of carbon black and silica. Two methods are employed to incorporate the fillers. In one case, carbon black and silica (as per the formulation in Table 2.8) are added to the latex as dispersions. In the second case, carbon black and silica in the same concetration, are incorporated to dry rubber by the mill-mixing process. The samples to which carbon black and silica are added in the latex stage are designated as wet mix and the samples to which carbon black and silica are added to dry rubber by the mill-mixing process are designated as dry mix.

Table 2.8 Formulation of Mixes for Latex-Stage Incorporation of Fillers (Latex Carbon Black Masterbatch)

Sample designation	Type of Mixing: Latex-Stage-Mixed Carbon Black Masterbatches				
	MBL1	MBL2	MBL3	MBL4	MBL5
40% Fresh Natural rubber Latex	250	250	250	250	250
20% Carbon black (ISAF) Dispersion	225	250	200	225	200
25% Precipitated silica Dispersion	0	0	20	20	40
20% Fatty acid soap solution	5	5	5	5	5
Total wet weight	480	505	480	505	505
Total dry weight	146	151	146	151	151

The formulations for stage 1 of wet stage mixing is given in Table 2.8. and for dry stage mixing of carbon black masterbatch is given in Table 2.9. A second stage mixing is performed on a two-roll mixing mill to add other compounding ingredients as per Table 2.10.

Table 2.9 Formulation of Mixes for Dry-Stage Incorporation of Fillers (Carbon Black Masterbatch)

Sample designation	Type of Mixing: Dry-Stage-Mixed Carbon Black Masterbatches				
	MBD1	MBD2	MBD3	MBD4	MBD5
Natural rubber	100	100	100	100	
Carbon black (ISAF)	45	50	40	45	40
Precipitated silica	0	0	5	5	10
Fatty acid	1	1	1	1	
Total weight	146	151	146	151	151

The extrudates obtained with a Brabender attached to a Garvey die head with a die of 3 mm and diameter 19 mm and L/D of 20 are shown in Figure 2.16 and Figure 2.17. For the assessment of extrudate performance, the following conditions were maintained: barrel temperature 70 °C, head temperature 100 °C, die temperature, 110 °C and screw speed 40 and 60 rpm.

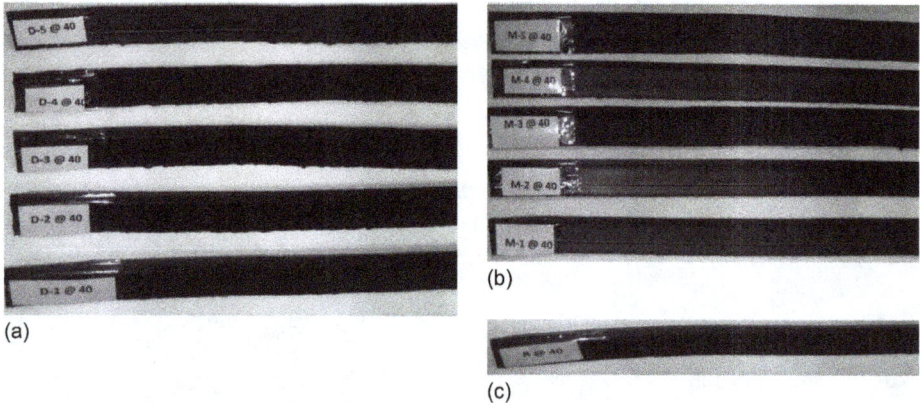

(a)

(b)

(c)

Figure 2.16 Extrudate appearance extruded at a screw rpm of 40 (a: dry mixed, b: wet mixed, c: regular (control mix)) (Courtesy: Apollo Tyres Ltd., Chalakudi, Kerala)

All the samples contain the same amount of carbon black, but the filler is mixed on a mixing mill in the case of samples denoted as dry and the filler is incorporated at the latex stage for those denoted as wet.

Table 2.10 Formulation for Second-Stage Mixing of the Carbon Black Masterbatches

Ingredients	Control	Dry-Stage Mixed					Latex-Stage Mixed				
	Regular (R)	D-1	D-2	D-3	D-4	D-5	M-1	M-2	M-3	M-4	M-5
Stage 1 masterbatch (MBD)	134	146	151	146	151	151	-	-	-	-	-
Stage 1 masterbatch (MBL)	-	-	-	-	-	-	146	151	146	151	151
Carbon black (ISAF)	25	3	0	8	3	8	3	0	8	3	8
Precipitated silica	0	8	8	3	3	0	8	8	3	3	0
Stearic acid	0	2	2	2	2	2	2	2	2	2	2
ZnO	4.5	4.5	4.5	4.5	4.5	4.5	4.5	4.5	4.5	4.5	4.5
DTPD*	0.5	0.5	0.5	0.5	0.5	0.5	0.5	0.5	0.5	0.5	0.5
6PPD	2.5	2.5	2.5	2.5	2.5	2.5	2.5	2.5	2.5	2.5	2.5
Wax	1.5	1.5	1.5	1.5	1.5	1.5	1.5	1.5	1.5	1.5	1.5
TMQ	0.75	0.75	0.75	0.75	0.75	0.75	0.75	0.75	0.75	0.75	0.75
Total weight	168.75	168.75	170.75	168.75	168.75	170.75	168.75	170.75	168.75	168.75	170.75

* DTPD- N,N'-Ditolyl-p-phenylenediamine

All the samples contain the same amount of carbon black, but the filler is mixed on a mixing mill in the samples denoted as dry and is incorporated in the latex stage in those denoted as wet. It can be seen that the surface is more uniform and smooth for wet mixes.

dry-mixed

wet-mixed

regular (control mix)

Figure 2.17 Appearance of extrudate extruded at a screw rpm of 60 (a: dry mixed, b: wet mixed, c: regular (control mix)) (Courtesy: Apollo Tyres Ltd., Chalakudi, Kerala)

2.9.4 Factors Affecting Extrudate Performance of Dry- and Latex-Stage Mixed Mixes

It is observed that the surface is more uniform and smoother for wet mixes and only marginally higher for dry mixes where carbon black is added by the mill-mixing process.

Carbon black is incorporated into fresh, natural rubber latex. Since the latex is not preserved, it has a lower gel content and the rubber molecules are long and branched. This is maintained to a certain degree, as mechanical mastication is performed only in the second stage of mixing, where other additives are added. The degree of filler dispersion is high, because carbon black is added as an aqueous dispersion.

It can be seen that, when carbon black is incorporated at the latex stage as a dispersion, the reduction in molecular weight is minimized. This is further supported by drying at 70 °C, which helps preserve the high molecular weight of the natural rubber. As a result, latex-stage incorporation leads to improved filler dispersion and enhanced elasticity.

Die swell measured with a Brabender extruder fitted with a Garvey die head is influenced by factors such as the rubber compound, die geometry and extrusion conditions. A longer relaxation time corresponds to a lower stress relaxation rate, resulting in a greater elastic response and, consequently, increased extrudate swell.

2.9.4.1 Die Parameters Affecting Die Swell

The different regions of the die that influence die swell are mainly related to the flow in the barrel, the die's entrance region and the die's exit region, as the rubber enters the die entrance region with a stress (σ). During shear flow in the die, some stress decay in the rubber chains as they exit from the Garvey head die. On exiting, the molecular recovery stage occurs. Some parts of elastic strain are recovered, due to the storage of elastic energy during shear deformation in the Garvey head die by the elastic component of rubber. The greater the stored elastic energy, the greater is the extrudate swell. Other factors such as the length of the barrel and the shear rate experienced during the flow inside the die also affect die swell.

2.9.4.2 Factors Influencing Rubber Flow and Die Swell

The stress decay that happens in the die depends on the rubber compound and the residence time in the capillary. The longer the residence time, the longer it takes the stressed molecules to dissipate their stored energy; hence, die swell is lower [40].

2.9.4.3 Effect of Micro- and Macrostructure on Rubber

In general, the higher the molecular weight, the better are the mechanical properties. Natural rubber has a very high molecular weight compared with synthetic rubbers.

The higher molecular weight and molecular entanglements can result in higher elasticity and, hence, greater die swell. A broad molecular weight distribution exhibits greater elasticity during shear flow, but a slower rate of elastic recovery than a narrow molecular weight distribution. Chain branching also causes increased die swell because the relaxation rate is lower than in linear polymers.

Regarding the filler effect, die swell increases with increase in filler dispersion. Due to filler deagglomeration, an effective reduction in filler volume occurs. The effective filler volume fraction increases if immobilized rubber increases.

Die swell increases with increase in total recoverable shear strain and elastic energy.

The advantages of latex-stage incorporation filler over dry-rubber incorporation of fillers are evident from extrudates.

Notations Used

$\dot{\gamma}$	Shear rate
η	Viscosity
k	Flow consistency index
n	Power-law index
τ	Shear stress
D	Die diameter
De	Extrudate diameter
L	Capillary length
R	Capillary radius

References

[1]   R. H. Colby, L. J. Fetters, W. W. Graessley, 'The melt viscosity-molecular weight relationship for linear polymers', *Macromolecules*, vol. 20, no. 9, pp. 2226–2237, Sep. 1987, doi: 10.1021/ma00175a030.

[2]   E. B. Bagley, 'The Separation of Elastic and Viscous Effects in Polymer Flow', *Trans. Soc. Rheol.*, vol. 5, no. 1, pp. 355–368, Mar. 1961, doi: 10.1122/1.548905.

[3]   A. V. Tobolsky, D. Katz, M. Takahashi, 'Rheology of polytetrafluoroethylene', *J. Polym. Sci. A*, vol. 1, no. 1, pp. 483–489, Jan. 1963, doi: 10.1002/pol.1963.100010142.

[4]   H. Leaderman, 'Elastic and creep properties of filamentous materials', Massachusetts Institute of Technology, Cambridge, 1943.

[5]   J. L. White, *Rubber processing: technology, materials, principles*. Munich ; New York ; Cincinnati: Hanser Publishers ; Hanser/Gardner Publications, 1995.

[6]   P. S. Johnson, *Rubber processing: an introduction*. München: Hanser, 2001.

[7]   C. M. Roland, *Viscoelastic Behavior of Rubbery Materials*. Oxford: Oxford University Press, Incorporated, 2011.

[8] H. C. Baker, R. M. Foden, 'Recent Developments in Superior Processing Natural Rubber', *Rubber Chem. Technol.*, vol. 33, no. 3, pp. 810–824, Jul. 1960, doi: 10.5254/1.3542199.

[9] B. C. Sekhar, G. W. Draske, 'Superior Processing Rubber Masterbatch', *J. Rubber Res. Inst. Malaya*, 1958.

[10] J. Leblanc, 'Rubber–filler interactions and rheological properties in filled compounds', Prog. Polym. Sci., vol. 27, no. 4, pp. 627–687, May 2002, doi: 10.1016/S0079-6700(01)00040-5.

[11] M. Morton, *Rubber technology*, 3rd ed. Dordrecht: Kluwer Academic Publishers, 1999.

[12] C. M. Blow, S. H. Morell, 'The chemistry and technology of vulcanization', in *Rubber technology and manufacture*, London: Butterworths for the Institution of the Rubber Industry, 1971.

[13] R. Alex, P. P. De, S. K. De, 'Self-vulcanizable ternary rubber blend based on epoxidized natural rubber, carboxylated nitrile rubber and polychloro-prene rubber: 2. Effect of blend ratio and fillers on properties', Polymer, vol. 32, no. 14, pp. 2546–2554, Jan. 1991, doi: 10.1016/0032-3861(91)90334-F.

[14] S. Mukhopadhyay, S. K. De, 'Miscibility of self-vulcanizable rubber blend based on epoxidized natural rubber and chlorosulphonated polyethylene: effect of blend composition, epoxy content of epoxidized natural rubber and reinforcing black filler', Polymer, vol. 32, no. 7, pp. 1223–1229, Jan. 1991, doi: 10.1016/0032-3861(91)90225-8.

[15] A. Limper, *Mixing of rubber compounds*. Munich : Cincinnati, Ohio: Hanser Publishers ; Hanser Publications, 2012.

[16] P. S. Johnson, 'Mixing Technology', in Rubber Products Manufacturing Technology, 1st ed., A. K. Bhowmick, M. M. Hall, H. A. Benarey, Eds., Routledge, 2018, pp. 103–121. doi: 10.1201/9780203740378-2.

[17] N. Nakajima, 'Elongational Flow in Mixing Elastomer with Carbon Black', *Rubber Chem. Technol.*, vol. 53, no. 5, pp. 1088–1105, Nov. 1980, doi: 10.5254/1.3535080.

[18] B. B. Boonstra, A. I. Medalia, 'Effect of Carbon Black Dispersion on the Mechanical Properties of Rubber Vulcanizates', *Rubber Chem. Technol.*, vol. 36, no. 1, pp. 115–142, Mar. 1963, doi: 10.5254/1.3539530.

[19] E. S. Dizon, 'Processing in an Internal Mixer as Affected by Carbon Black Properties', *Rubber Chem. Technol.*, vol. 49, no. 1, pp. 12–27, Mar. 1976, doi: 10.5254/1.3534941.

[20] J. W. ten Brinke, S. C. Debnath, L. A. E. M. Reuvekamp, J. W. M. Noordermeer, 'Mechanistic aspects of the role of coupling agents in silica–rubber composites', Compos. Sci. Technol., vol. 63, no. 8, pp. 1165–1174, Jun. 2003, doi: 10.1016/S0266-3538(03)00077-0.

[21] W. Kaewsakul, K. Sahakaro, W. K. Dierkes, J. W. M. Noordermeer, 'Optimization of Mixing Conditions for Silica-Reinforced Natural Rubber Tire Tread Compounds', *Rubber Chem. Technol.*, vol. 85, no. 2, pp. 277–294, Jun. 2012, doi: 10.5254/rct.12.88935.

[22] R. Alex, K. K. Sasidharan, T. Kurian, A. K. Chandra, 'Carbon black masterbatch from fresh natural rubber latex', *Plast. Rubber Compos.*, vol. 40, no. 8, pp. 420–424, Oct. 2011, doi: 10.1179/1743289810Y.0000000038.

[23] A. K. Bhowmick, M. M. Hall, H. A. Benarey, Eds., *Rubber products manufacturing technology*. New York: Dekker, 1994.

[24] S. Kawahara, T. Kakubo, J. T. Sakdapipanich, Y. Isono, Y. Tanaka, 'Characterization of fatty acids linked to natural rubber—role of linked fatty acids on crystallization of the rubber', Polymer, vol. 41, no. 20, pp. 7483–7488, Sep. 2000, doi: 10.1016/S0032-3861(00)00098-7.

[25] Y. Nie, Z. Gu, Y. Wei, T. Hao, Z. Zhou, 'Features of strain-induced crystallization of natural rubber revealed by experiments and simulations', *Polym. J.*, vol. 49, no. 3, pp. 309–317, Mar. 2017, doi: 10.1038/pj.2016.114.

[26] S. Amnuaypornsri, S. Toki, B. S. Hsiao, J. Sakdapipanich, 'The effects of endlinking network and entanglement to stress–strain relation and strain-induced crystallization of un-vulcanized and vulcanized natural rubber', *Polymer*, vol. 53, no. 15, pp. 3325–3330, Jul. 2012, doi: 10.1016/j.polymer.2012.05.020.

[27] N. Nishiyama, S. Kawahara, T. Kakubo, E. A. Hwee, Y. Tanaka, 'Origin of Characteristic Properties of Natural Rubber—Synergistic Effect of Fatty Acids on Crystallization of cis-1,4-Polyisoprene: II,

Mixed and Esterified Fatty Acids in Natural Rubber', *Rubber Chem. Technol.*, vol. 69, no. 4, pp. 608–614, Sep. 1996, doi: 10.5254/1.3538388.

[28] B. Huneau, 'Strain-Induced Crystallization of Natural Rubber: A Review of X-Ray Diffraction Investigations', *Rubber Chem. Technol.*, vol. 84, no. 3, pp. 425–452, Sep. 2011, doi: 10.5254/1.3601131.

[29] J. A. Cruz-Morales, C. Gutiérrez-Flores, D. Zárate-Saldaña, M. Burelo, H. García-Ortega, S. Gutiérrez, 'Synthetic Polyisoprene Rubber as a Mimic of Natural Rubber: Recent Advances on Synthesis, Nanocomposites and Applications', *Polymers*, vol. 15, no. 20, p. 4074, Oct. 2023, doi: 10.3390/polym15204074.

[30] J. Tangpakdee, Y. Tanaka, 'Characterization of Sol and Gel in Hevea Natural Rubber', *Rubber Chem. Technol.*, vol. 70, no. 5, pp. 707–713, Nov. 1997, doi: 10.5254/1.3538454.

[31] N. Ichikawa, E. A. H, T. Y, 'Properties of Deproteimzed Natural Rubber Latex', presented at the Int Rubb Technol Conf, Kuala Lumpur, pp. 11–20.

[32] S. Liao, 'A Review on Characterization of Molecular Structure of Natural Rubber', *MOJ Polym. Sci.*, vol. 1, no. 6, Dec. 2017, doi: 10.15406/mojps.2017.01.00032.

[33] P. Rojruthai, T. Kantaram, J. Sakdapipanich, 'Impact of non-rubber component on the branching structure and the accelerated storage hardening in Hevea rubber', J. Rubber Res., vol. 23, no. 4, pp. 353–364, Dec. 2020, doi: 10.1007/s42464-020-00063-7.

[34] L. L. Blyler, 'Some Aspects of the Relationship between Molecular Structure and Flow Behavior of Polymer Melts', *Rubber Chem. Technol.*, vol. 42, no. 3, pp. 823–834, Jul. 1969, doi: 10.5254/1.3539261.

[35] S.-N. Gan, 'Storage Hardening of Natural Rubber', *J. Macromol. Sci. Part A*, vol. 33, no. 12, pp. 1939–1948, Dec. 1996, doi: 10.1080/10601329608011018.

[36] R. I. Wood, 'Mooney Viscosity Changes in Freshly prepared -Raw Natural Rubber', *J Rubb Res Inst Malaya*, vol. 14,20.

[37] M. D. Morris, 'Contribution of Storage Hardening to Plasticity Retention Index Test for Natural Rubber', *J Nat Rubb Res*, vol. 6(2), pp. 96–104.

[38] *Standard Test Methods for RubberĐViscosity, Stress Relaxation and Pre-Vulcanization Characteristics (Mooney Viscometer)*, 1646, 2004.

[39] E. B. Bagley, 'The Separation of Elastic and Viscous Effects in Polymer Flow', *Trans. Soc. Rheol.*, vol. 5, no. 1, pp. 355–368, Mar. 1961, doi: 10.1122/1.548905.

[40] C. Sirisinha, 'A Review of Extrudate Swell in Polymers', *ScienceAsia*, vol. 23, no. 4, p. 259, 1997, doi: 10.2306/scienceasia1513-1874.1997.23.259.

3
Viscoelasticity of Rubber

Rosamma Alex, Pradeepkumar P. Joy, Umasankar G

3.1 Introduction

Rubber attains its utility as an industrial product only after undergoing vulcanization. The presence of reinforcing fillers strongly affects its viscoelastic properties. Moreover, the method of by which fillers are incorporated plays a significant role in filler dispersion, which in turn affects the viscoelastic behavior of the vulcanized rubber.

It is assumed that rubber molecules are highly flexible because of the facile rotation of carbon-carbon bonds that form the backbone of the rubber chain and the ability to undergo conformational changes without affecting internal energy. The rates of these entropic changes are governed mainly by molecular structure and temperature. The dependence of conformational change on temperature is based on the Arrhenius equation for free volume, as free volume increases with temperature above the glass transition temperature (T_g) of rubber. At ambient temperatures, the rates of conformational changes are so fast that they are essentially instantaneous, even under very high strain. The rubber-like response happens at ambient temperatures and the relaxation involves the whole molecular chain as a unit. However, near the T_g, the response rates are very low, resembling a glass-like behavior. In this region, bond stretching and relaxation processes involving small segments of the rubber chains are minimal due to insufficient thermal energy and the close packing of the chains. The mobility of rubber molecules can be related to the space available for conformational changes, called the free volume of the rubber. The mobility can also be affected by factors such as chain entanglements in the vulcanized network, free rubber chain segments, and some side chains, due to the branching of rubber molecules, rubber filler interactions and filler interactions.

When a rubber sample is exposed to stress or strain, the rubber molecules undergo internal rearrangement, such as rotation and extension of the polymer chains. Internal rearrangement is influenced by interactions between rubber chains themselves or between rubber chains and fillers. During deformation, weak interactions involving rubber chains or rubber chains and filler may be ruptured and disentanglements may occur, resulting in loss of energy.

This loss of energy during deformation imparts a viscous nature to rubber, because internal friction is generated in the rubber matrix. Hence, rubber exhibits viscoelastic properties in dynamic conditions.

The viscoelasticity of a vulcanized filler rubber is a relaxation process involving a decrease in stress over time during deformation. Four viscoelastic parameters influenced by filler type and dispersion in filled rubber vulcanizate are

- Stress relaxation

- Creep

- Hysteresis

- Phase lag between stress and strain in sinusoidal deformation of rubber

Stress relaxation is the decay in stress over time of samples subjected to constant strain. In typical stress relaxation tests involving high strain, the initial stress or relaxation modulus is high, but decreases progressively with time. The initially high modulus reflects a glassy nature, while the low modulus observed corresponds to a rubbery response.

Stress relaxation and creep tests provide information about the response of rubber molecules to stress over both short and long periods of time. However, for practical purposes, the tests are used to determine long-term performance in components such as seals and gaskets. For evaluating short-term responses, sinusoidal deformation tests are employed, especially for products such as engine mounts, vibration absorbers and tires.

In sinusoidal deformation, the rubber molecules do not have sufficient time to relax to an equilibrium condition, especially at very high deformation frequencies; relaxation happens rapidly and corresponds to a glassy response. This can lead to an equivalence between time (deformation frequency) and temperature, where increase in temperature or decrease in deformation frequency leads to a transition from glassy to rubbery behavior. This time–temperature equivalence has practical significance for predicting the properties of rubber products.

The mathematical expression of viscoelastic behavior at a particular temperature is based on a spring and dashpot arranged in series (Maxwell model) or in parallel (Voight model). Stress relaxation is the decrease in stress that happens exponentially in a sample maintained at constant deformation and is explained by the Maxwell model. Creep behavior and relaxation under sinusoidal deformation over a wide range of frequencies is explained by the Voigt model.

The mechanical properties of rubber are improved by the addition of fillers, which are classified mainly as reinforcing, semi-reinforcing and non-reinforcing. Reinforcing fillers, such as carbon black and silica, considerably improve mechanical properties, such as tensile strength, tear strength, abrasion resistance and modulus. However, the viscoelastic response is enhanced, as evidenced by increased stress relaxation, creep and energy loss. In carbon-black-filled vulcanizates, the stress relaxation rate is higher than in gum rubber, as stress relaxation involves rubber–filler and filler–filler interactions involving carbon black.

3.2 Brief History of Rubber

The earliest observations of rubber use are credited to Christopher Columbus and his crew during their voyages to the Caribbean, where they observed people playing with rubber-like balls. In 1524, such rubber balls were presented at the court of Emperor Charles V of Spain. Later, in 1751, a French explorer, Jean Marie de la Condamine described to the Royal Academy of Sciences how Amazonian natives extracted milky exudates from forest trees to produce items like shoe soles and waterproof fabrics. In 1770, Joseph Priestly named this material 'rubber' because it could remove pencil marks. In French, the name 'caoutchouc' was derived from the native Amazonian expression 'caa-o-chu,' which means 'weeping tree,' and remains in common parlance in French. There are reports that, in 1823, Charles Macintosh of Scotland produced waterproof clothes using rubber in solvent naphtha [1, 2].

In 1820, Thomas Hancock invented the rubber masticator, a machine for incorporating different chemicals into the rubber. A major breakthrough followed with the discovery of sulfur vulcanization. In 1839, Charles Goodyear in the United States accidentally dropped a mixture of Indian rubber and sulfur on a hot stove and observed that the material acquired a different nature. Although this discovery was accidental, both Goodyear (on June 15) and Thomas Hancock (May 21) independently developed and patented the vulcanization process in 1844. This innovation, emerging during the Industrial Revolution marked the transformation of rubber into a vital raw material for the rubber industry.

The pioneer of the modern rubber industry is considered to be John Boyd Dunlop, a Scots-Irish physician who invented a pneumatic tire for his son's cycle. The story goes that his son won the race on a tricycle fitted with a rubber 'tyre', which he patented in 1888. Another milestone was the use of fillers, especially carbon black [1, 3]. It was only later that the tremendous property enhancement by carbon black was realized and it is now an indispensable filler for the tire industry.

Before that, latex had become a precious commodity and trees had been planted. In Singapore in 1877, the British planted rubber from seeds that had been collected at Kew Gardens in England. Similar plantations arose in British Malaysia, French Indochina and the Dutch East Indies. By 1910, plantation production started to exceed demand. Later, the supply of rubber was limited in many countries. During World War I, Germany developed a synthetic rubber, Buna N, from butadiene. In World War II, the United States intensified its production of synthetic rubber after Japan seized control of rubber-producing areas of southeast Asia.

3.3 Natural Rubber

Natural rubber is a high-molecular polymer made from the monomer isoprene and chemically is cis-1,4-polyisoprene. The polymer molecules have almost negligible intermolecular interactions and so the molecules are free to vibrate and perform random movements at ambient temperature. There can be interactions involving associated proteins and linked phospholipids of rubber molecules that weakly crosslink the rubber chains. Natural rubber is an amorphous polymer whose chain flexibility is greatly related to temperature. At ambient temperatures, the molecules tend to be randomly coiled rather than straight or uncurled, as entropy in coiled form is much greater than in the straightened position because carbon–carbon single bonds have great freedom to rotate. There is no bond length or angle change in these randomly coiled forms and no change in internal bonding energy. When the molecules are stretched, they can simply slide past one another. If some physical or chemical cross-links are introduced, then when these molecules are pulled, parts of the interior become extended. As the temperature increases, this coiling tendency becomes more pronounced. Consequently, the force needed to stretch the molecules to a particular extent also increases. As the temperature decreases, a temperature is reached where segmental movement almost stops, which is called the glass transition temperature (T_g). Usually, segmental mobility is of two types: an elastic solid and a viscous fluid. The segmental motion of the rubber molecules above T_g is the main factor responsible for the key characteristics of a polymer. In rubber, a lower glass transition temperature corresponds to higher chain flexibility at room temperature and a more pronounced rubbery nature.

3.4 Vulcanization of Rubber and its Mechanical Properties

The specific physical properties, such as viscoelasticity, resilience, impact and abrasion resistance, easy processing characteristics and green strength, make NR an important raw material in the manufacture of many different rubber and latex products. The full potential of these properties is realized when the rubber molecules are cross-linked. The basic process involved in making raw rubber a useful rubber in product manufacture is vulcanization, which transforms raw rubber into a highly elastic industrial raw material. Vulcanization is influenced by the additives employed in rubber compounding. Acidic additives can adversely affect the rate and extent of vulcanization. They include acidic fillers, such as silica, clays, channel-grade carbon black and acidic processing aids, such as pine tar. Fillers such as carbon black are known to influence vulcanization by increasing rheometric torque. Hence, it is important to understand the role and type of fillers, their method of incorporation and their impact on the viscoelastic properties of vulcanized natural rubber [4].

All the mechanical and dynamic properties related to modulus, strength, elongation and resilience improve as the cross-link density increases [4, 5]. This is shown schematically in Figure 3.1. Typically, the maximum tear strength occurs at a lower cross-link density than the maximum tensile strength. Generally, as modulus increases, elongation at break decreases. Both static and dynamic modulus increase as cross-link density increases. The vulcanizate properties also depend on the type of cross-links. Generally, polysulfide cross-links yield better mechanical properties, while mono-sulfidic and carbon–carbon cross-links confer better aging properties and lower compression set.

The average molecular weight of the polymer between two adjacent cross-links (Mc) can be related to the vulcanizate's tensile strength, which can be measured from values of the volume fraction of rubber in a swollen vulcanizate in a suitable solvent (V_r). The cross-link density can be calculated with the Flory–Rehner Equation (Equation 3.1), using the volume fraction of rubber in the swollen sample. This approach is based on investigations carried out by Paul Flory and J. Rehner in the early 1940s [6]. They developed a model that describes the isotropic swelling of rubber cross-linked in the dry state. Mc is the molecular weight of the rubber between cross-links.

The volume fraction of rubber (V_r) was calculated as follows from the above measurement:

$$V_r = \frac{(D - FT)/\rho_r}{((D - FT)/\rho_r) + (A_0/\rho_s)} \qquad (3.1)$$

Where

T = Weight of the test specimen

D = Deswollen weight

F = Weight fraction of insoluble components

A_0 = Weight of the absorbed solvent, corrected for the swelling increment

ρ_r and ρ_s = Densities of rubber and solvent, respectively

Cross-link density is calculated as per the Flory–Rehner equation given below:

$$v = \frac{-\ln(1 - V_r) + V_r + \chi V_r^2}{V_s \left(V_r^{\frac{1}{3}} - 0.5V_r\right)} \tag{3.2}$$

Where

v = cross-link density

V_r = volume fraction of rubber in equilibrium-swelled vulcanizate sample

V_s = molar volume of used solvent

χ = Huggins parameter

n = number of network chain segments bounded on both ends by cross-links

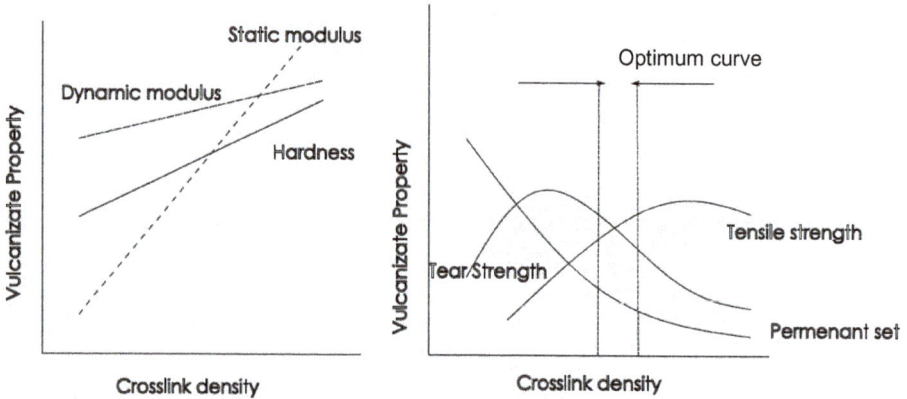

Figure 3.1 Schematic representation of the dependence of properties of rubber on cross-link density

3.5 Fillers and Mechanical Properties of Rubber Vulcanizates

The mechanical properties of rubber are improved by the incorporation of fine particulate fillers, such as silica and carbon black, into the rubber matrix [7]. The improvement in properties such as tensile strength, tear strength, abrasion resistance and modulus by fillers is referred to as rubber reinforcement and the fillers that impart reinforcement are called reinforcing fillers. Based on the level of reinforcement, fillers are basically referred to as reinforcing, semi-reinforcing and non-reinforcing. While reinforcing fillers considerably improve tensile strength, tear strength and abrasion resistance, semi-reinforcing fillers improve these parameters only slightly, and non-reinforcing fillers have no appreciable effect on them. Non-reinforcing fillers are used for play balls and the like, because they do not adversely affect dynamic properties, such as resilience.

Along with vulcanization, reinforcement by fillers such as carbon black and silica, which are the two key fillers, controls properties such as modulus hysteresis and also failure properties such as wear resistance, tear strength, and cut growth resistance. One definition of rubber reinforcement is the improvement in the mechanical properties in the rubber matrix, such as tensile strength, tear strength, abrasion resistance and modulus. Reinforcement depends on the level of vulcanization or cross-link density of the rubber matrix.

Reinforcement can be mathematically described as a hydrodynamic effect that leads to an increase in viscosity. An increase in fluid viscosity, due to dispersed rigid particles (mainly fillers, such as carbon black), was introduced as a hydrodynamic effect by Einstein [8, 9] because rigid bodies decrease the velocity gradient during flow, leading to an increase in viscosity.

How the viscosity increases was modeled by the following equation:

$$\eta = \eta_0(1 + 2.5\varphi) \tag{3.3}$$

Where

φ = filler volume fraction

η_0 = viscosity of the pure fluid without fillers

A parameter that is likely to obey this model in vulcanized elastomers is modulus. Based on Young's modulus, E, the equation can be modified as:

$$E = E_0(1 + 2.5\,\varphi) \tag{3.4}$$

Where

E and E_0 = Young's moduli of filled and unfilled elastomers

φ = volume fraction of filler

The equation was modified to allow for higher filler concentrations by considering the hydrodynamic interactions between pairs of particles [10].

$$E = E_0(1 + 2.5\,\varphi + 14.1\,\varphi^2) \tag{3.5}$$

Another term, the ratio of E to E_0, was coined to describe the effect of filler on elastomer strain.

The ratio E/E_0 is called the strain amplification factor and indicates the local strain experienced by the rubber chains in the presence of rigid fillers, which themselves do not undergo deformation during extension of the elastomer matrix.

However, for reinforcing fillers that show rubber–filler interaction (and when used in high proportions for practical applications), the modulus observed in reinforced carbon-black-filled matrix is higher than that predicted by the equations cited (Equation 3.5). Consequently, an equation was proposed to describe carbon black filler aggregates dispersed within a rubber matrix, incorporating a factor f that accounts for the characteristics of the filler aggregates.

$$E = E_0(1 + 0.67f\varphi + 1.62f^2\varphi^2) \tag{3.6}$$

The equation assumes that there is no chemical interaction between filler and rubber.

It has been observed that incorporating and dispersing carbon black into rubber can increase its modulus more than tenfold compared with gum vulcanizate, without adversely affecting other properties, such as strength and elongation.

3.5.1 Significance of Carbon Black

Carbon black is a pure form of elemental carbon comprising microscopic, mostly spherical particles that exist as aggregates of carbon black particles; carbon black is about 97–99% elemental carbon with 1–3% hydrogen, sulfur, nitrogen and oxygen present as functional groups on the carbon black surface. This chemical composition influences carbon black's morphology, surface activity and reinforcement characteristics. Carbon black particles are partly graphitic in structure. The layers of carbon atoms that form graphitic structures are arranged systematically and also randomly. Due to this random orientation of layers of carbon atoms, the carbon atoms at the layer edges can have unsaturated carbon bonds that provide sites for chemical reactivity.

Carbon black exists as aggregates of carbon black particles and, due to the presence of imperfect graphitic layers, carbon atoms at exposed edges have specific functional groups, such as anhydride, aldehyde, lactone, phenol, carboxyl, carboxylate etc., that provide a site for chemical interaction with rubber matrix. Chemical compatibility between carbon black and natural rubber is a driving force that enables rubber reaction with active sites on a carbon black surface, leading to good polymer–filler interaction.

The surface area of a filler and carbon black particularly indicates the size of the contact area between the rubber and the filler surface. Generally, carbon black with a small particle size may have a high external surface area and the morphology of the carbon black aggregates can influence this. Generally, carbon black with a small particle size may have a high external surface area and a higher reinforcement potential. The morphology of the carbon black aggregates can also influence this [7].

The aggregates of carbon black particles fuse together as agglomerates, which break up when mixed with rubber. The aggregates are regarded as the smallest dispersible unit of carbon black within the rubber matrix in typical conditions. As a result, aggregates are often considered discrete units of carbon black and the actual reinforcing entity [7].

The average primary particle diameter of furnace blacks varies from 17–110 nm. Particle size and distribution can be considered the key parameters for the reinforcement capacity of carbon black.

How carbon black fuses to form aggregates is defined by the term structure or three-dimensional form of carbon black particles. In a rubber compound, these void spaces can be filled by rubber. This rubber is called occluded rubber [11], is partially shielded from deformation and acts as a part of the filler. High-structure carbon black can have high reinforcing potential.

The rubber chains can interact with fillers, which is considered a physical interaction and also by covalent bonds to an extent depending on the surface activity of carbon black [12, 13]. As a result, bound rubber will form in the rubber compound before vulcanization and can be quantified by its resistance to dissolution in a good rubber solvent.

The reinforcement of rubber compounds by carbon black is due to weaker interactions between the carbon black surface and rubber molecules *via* physical adsorption and mechanical interlocking by virtue of small particle size, high surface area, structure and functional groups of carbon black aggregates. All these lead to an improved rubber–filler interaction, as seen in the formation-bound rubber. In other words, bound rubber formation results from a combination of physical adsorption, chemisorption and mechanical interlocking. Carbon black is a highly engineered nanofiller and the reinforcement characteristics of carbon-black-filled rubber are highly complex. However, better filler dispersion that enables better rubber–filler interactions can remain crucial in imparting reinforcement by carbon black.

The reinforcing effect of carbon black needs to be understood in terms of the mechanical properties of the vulcanized rubber.

Carbon black is mainly produced by the incomplete combustion of hydrocarbons and also by thermal cracking (vapor-phase pyrolysis). The major types of carbon black are furnace black (produced by the oil furnace process), channel black (impingement on channels), thermal black (by decomposition of natural gas in the absence of oxygen), acetylene black (thermal decomposition of acetylene gas in the absence of oxygen) and lamp black (by burning oil in a limited supply of oxygen). In the furnace process for carbon black production, a hydrocarbon feedstock is injected into a high-temperature combustion zone, where it undergoes incomplete combustion under a controlled supply of oxygen; the resulting material is then quenched with water to yield carbon black as a fine powder. The combustion reaction occurs in less than a second, during which time the nanosized carbon black particles fuse *via* covalent bonds to form aggregates. Over the next few seconds, the aggregates fuse to form agglomerates through electrostatic bonds that can be rupture during processing on a mixing mill or other mixing equipment.

The various grades of furnace black are super-abrasion furnace black (SAF), intermediate super-abrasion furnace (ISAF), high-abrasion furnace (HAF), fast-extruding furnace black (FEF), general-purpose furnace black (GPF) and semi-reinforcing furnace (FEF). Thermal blacks are produced by thermal decomposition at 1300 °C in the absence of oxygen in cylindrical furnaces lined with silica bricks. The grades are fine-thermal (FT) and medium-thermal (MT). Lamp black is made by burning petroleum or coal tar residues. Acetylene black is separated from hot gas streams and smoke by suitable cooling and pelletizing process. Acetylene black, as the name suggests is produced by the thermal decomposition of acetylene gas and during this process carbon particles arrange in a specific manner to have structure which contributes to its property of high conductivity.

The extent of vulcanization and the type of cross-links depend not only on the type of vulcanizing chemicals, but also on reinforcing fillers, such as carbon black, and hence the mechanism of vulcanization [14]. It is generally observed that vulcanization initiates earlier and that the rate and level of vulcanization are higher in the presence of carbon black and loading of carbon black, as indicated by rheometric torque values (measured with a rheometer), which is a measure of cross-linking. Reports show that carbon black, by virtue of its specific chemical nature, can adsorb accelerator intermediates on its surface, thereby facilitating sulfur-based cross-linking, possibly involving mono-sulfidic cross-linkages [15, 16].

Generally, an increase in stiffness-related properties, such as higher tear strength and abrasion resistance, are considered the criteria for evaluating filler reinforcement. The strength parameters are improved by an enhanced level of vulcanization and the type of cross-links. A common measure of reinforcement is the ratio of the 300% modulus to the 100% modulus, the reinforcement index, which is derived from stress–strain curves. This reinforcement index tends to improve with increase in mixing time, the assumption being that the rubber–filler interaction based on bound-rubber formation is enhanced by the shear forces exerted in the mixing process and that aggregates could become fractured by shear forces [17].

3.5.2 Significance of Silica

Silica is produced by acidification of sodium silicate under controlled conditions. In the rubber industry, two types of silica are commonly used as reinforcing fillers: precipitated silica and fumed silica. Precipitated silica is amorphous, consisting of very small particles that fuse together to form larger structures called aggregates. Aggregates can contain hundreds of primary particles and are assumed to be effective units for reinforcing rubber compounds. Like carbon black, silica is characterized according to its particle size and distribution, structure and surface chemistry. Various particle arrangements are possible when silica aggregates are formed. The structure of silica can be measured by the oil absorption method. A silica primary particle can be 20 nm in size, but, due to hydrogen bonding by the polar silanol groups on the surface or other interactions involving dipole–dipole forces and induced dipole forces, silica primary particles remain as aggregates or even agglomerates. Aggregates can have dimensions of 50–500 nm [18]. The filler–filler interactions can result in increased viscosity of the rubber matrix.

Silica filler, unlike carbon black, has poor compatibility with natural rubber, because of the highly polar functional groups on the silica surface. Conventional mixing of silica with natural rubber generally results in poor filler dispersion and inferior mechanical properties. It has been shown that the silica surface contains many siloxane and silanol groups in different configurations. There can be one hydroxyl group on a single silicone atom (isolated), two hydroxyl groups on adjacent silicone atoms (vicinal) or two hydroxyl groups on the same carbon atom (germinal). The compatibility with rubber and better filler dispersion are achieved with the aid of silane-coupling agents, bifunctional organosilicon compounds and a suitable mixing process [19]. The silane-coupling agents have the general formula X_3SiRY, where RY represents the organofunctional group mercaptan or disulfide with a short alkyl link with a silicon atom and X represents the alkoxy group. Two silane-coupling agents generally employed in the rubber industry are bis(triethoxysilylpropyl) disulfide (TESPD) and bis(triethoxysilylpropyl) tetrasulfide (TESPT). The bifunctional groups on TESPT molecules, i.e. ethoxy and tetrasulfide, can react with the silica surface and NR during mixing and vulcanization. The reinforcement of silica-based rubber compounds mainly stems from chemical reactions of silica and rubber with silane-coupling agents. For carbon black, the reinforcement mainly entails weak interactions rather than chemical reactions. Mixing temperature and mixing equipment are chosen such that interaction can occur between silica and rubber during the mixing stage. Filler dispersion and mainly chemical interactions driven by silanol groups and other interactions, small particle size, high surface area and the structure of silica aggregates can also contribute to reinforcement.

Silica, compared with carbon black, is characterized by weaker filler–polymer interactions and stronger filler–filler interactions. This results in a higher compound viscosity.

However, the combination of silica with a coupling agent leads to a higher reinforcing effect and confers different dynamic mechanical properties compared with carbon black.

From a technological point of view, there are problems with basic incompatibility and difficulty in the dispersion of fillers, such as reinforcing carbon black and silica. Mixing carbon black and silica as dispersions at the latex stage should solve some of these problems.

Incorporating fillers, carbon black and silica at the latex stage provides options for better filler dispersion of single or mixed fillers (without the use of coupling agents) and for retaining the higher molecular weight characteristics of the raw rubber.

As the loading of carbon black increases, mechanical properties such as strength and abrasion resistance improve, but the improvement is not seen above 50 parts per hundred rubber (phr) loading. The strength and tear resistance decrease after about 70 phr. Properties such as hardness, rolling resistance, heat build-up and wet-skid resistance increase with filler loading up to a concentration of about 90 phr.

3.6 Mechanical Properties Related to the Viscoelasticity of Rubber

Rubber molecules are flexible, due to facile rotation around carbon–carbon bonds. This flexibility allows the molecules to extend in the direction of applied stress, resulting in a decrease in conformational entropy. The deformations of elastic rubber are caused by a decrease in conformational entropy with very little change in their internal energy.

Rubber elasticity is due to conformational or entropic changes happening in rubber molecules during deformation. If a rubber band is stretched by force and maintained in the deformed condition, it will be observed that retractive force will decrease when being heated and the rubber band tends to contract [42]. This observation was made by Gouch in 1805 and came to be called as thermoelastic effect and is due to the ability of rubber chains to be in a state of increased entropy, as kinetic energy achieved by the rubber molecules increases, due to enhanced thermal agitation in the network [20].

This increase in the rate of conformational change with temperature can be based on free volume and modeled using an Arrhenius-type equation. The conformational changes are considered instantaneous at ambient temperature. As the temperature rise above T_g, the rate of conformational changes increases and hence the elastic modulus decreases. This molecular mechanism is responsible for the rubber-like elasticity.

Rubber-like elasticity allows a rubber to deform under stress and to fully return to its original shape after removal of the stress. But rubbers are viscoelastic by nature and so have fluid-like properties along with elasticity. When deformed by stress and maintained at the same stress, they can deform further such that, after removal of the stress, the rubber does not return to its original shape .

3.7 Viscoelastic Parameters

Four viscoelastic parameters of rubber related to vulcanization, type and amount of fillers and hence filler dispersion are (1) stress relaxation, (2) creep, (3) hysteresis and (4) phase lag between stress and strain in the sinusoidal deformation of rubber.

Stress relaxation, hysteresis and stiffness are enhanced when there is a high degree of reinforcement, due to the specific nature of rubbers and the presence of fillers. The dependence of these properties on strain history is called the Mullins effect. The underlying mechanism is molecular slippage at the interface between rubber molecular chains and the fillers [21]. The slippage of molecular chains can redistribute stress, a fact which helps prevent molecular rupture, but adversely affects dynamic properties, such as heat build-up and resilience.

Stress relaxation and creep tests are commonly used to evaluate or predict properties in products such as gaskets, seals and tubes. Sinusoidal deformation tests are employed for short-term response predictions, such as in engine mounts or shock absorbers.

3.7.1 Stress Relaxation and Creep

Stress relaxation refers to the continued decay of stress over time under a constant deformation, while creep refers to continued deformation over time under constant stress. These two parameters are crucial, because many rubber products are designed for use under compression and undergo prolonged deformation at constant stress or stress decay at constant strain. Therefore, rubber products should be designed to manage or minimize stress relaxation and creep in line with their specific applications.

A viscoelastic polymer, when subjected to deforming stress, undergoes strain and, if the strain is kept constant, there occurs an internal rearrangement of polymer chains or conformational changes to a system of equilibrium during deformation and in the deformed condition. This is referred to as relaxation. This process varies with temperature, the amount of stress applied, the duration and the stress-application rate. Stress relaxation increases with rise in temperature, higher strain levels and longer durations of applied strain. Stress relaxation modulus at low strain is more elastic, whereas it is primarily viscous at high strains. Aging of polymer can result in a more relaxation process.

There can also be chemical changes that lead to chemical stress relaxation. Physical stress relaxation occurs at lower temperatures and lower durations of stress relaxation. Chemical stress relaxation predominates at higher temperatures and longer duration of stress and can be due to oxidation reactions and scission of rubber chains.

Stress relaxation can be expressed as a percentage of stress decay at a particular time.

Stress relaxation at higher temperatures results in higher permanent set, as there is also the effect of rubber aging at high temperatures:

$$\text{Stress loss } (t), \% = \frac{\sigma_0 - \sigma_t}{\sigma_0} \times 100 \tag{3.7}$$

Where

σ_0 = initial stress

σ_t = stress at time t

Creep refers to continued deformation under constant stress. The length of a specimen maintained at constant strain increases over time. Creep at a time t can be expressed as its length at time t:

$$\text{Creep } (t), \% = \frac{l_t - l_1}{l_1} \times 100 \tag{3.8}$$

The residual deformation in samples experiencing stress relaxation and creep under
tensile load, after removal of the force, is called tension set:

$$\text{Set } (t)100\% = \frac{l_x - l_0}{l_i - l_0} \times 100 \tag{3.9}$$

Where

l_x = recovered length

l_o = initial length

l_i = extended length

If the sample is subjected to a compression mode of stress relaxation, then the residual deformation is expressed as compression set. If the sample is subjected to a cyclic
deformation at a constant strain, the stress can decrease over time. This phenomenon
is known as cyclic stress relaxation, and the irrecoverable deformation that remains
after the load is removed is referred to as permanent set

Stress relaxation experiments can be carried out at different rates of strain. When the
tests are done at high strain, the stress relaxation modulus, which is the ratio of stress
decay to strain, is high initially, but declines as time increases. In terms of the relaxation process, the high initial relaxation modulus observed can be related to the
glassy nature, while the low relaxation modulus can be related to the rubbery nature
of the rubber vulcanizate.

3.7.1.1 Stress Relaxation and Permanent Set

Stress relaxation is the stress reduction observed in a sample subjected to constant deformation (ε_0). It can be defined as stress decay or stress loss at a time t as a percentage:

$$\text{Stress decay } (t) = \frac{\sigma_0 - \sigma_t}{\sigma_0} \times 100 \tag{3.10}$$

Where

σ_0 = initial stress

σ_t = stress at a time t

$$\text{Relaxation modulus } E_R = \sigma_t / \varepsilon_0 \tag{3.11}$$

The stress relaxation per unit time is the rate of stress relaxation. Stress relaxation is believed to stem from the slipping of rubber chain entanglements that loosen the network of molecular chains, such that they apply less force. When natural rubber is exposed at elevated temperatures, simultaneous oxidation and scission reactions occur as a result of interaction with molecular oxygen. It is known that the cross-linking and scission reactions are responsible for the permanent set, which happens in rubber samples that are deformed and then relaxed after a certain period [22, 23]. Andrews, Tobolsky and Hanson [24] proposed a molecular theory for permanent set in rubber during stress relaxation at elevated temperatures. The permanent set was also related to creep phenomena observed in elastomers. The occurrence of permanent set is a serious limitation on the performance of rubber products, especially for gaskets and seals.

Changes in the network structure at the molecular level can be studed *via* continuous and intermittent stress relaxation at elevated temperatures. This process separates the degradative process into cross-linking and scission reactions. For a homogeneous network in which all network chains are at equilibrium length l, the equation relating stress and elongation to attain length l_u in terms of concentration of network chains in rubber is given by:

$$f = skT[(l/l_u)^2 - (l/l_u)] \tag{3.12}$$

Where

f = stress per unit of cross-sectional area

s = number of network chains per unit volume of rubber

k = Boltzmann constant and T is the absolute temperature

Thus, at a fixed extension and constant temperature, the stress produced is entirely due to the load sharing of the rubber chains (this is in the absence of filler. If filler is present that has rubber–filler interactions, then filler also takes part in the load-bearing process). Hence, if network chains break, stress decay would occur; hence at a time t:

$$f(t) \propto s(t) \tag{3.13}$$

$$f(t)/f(0) \propto s(t)/s(0) \tag{3.14}$$

Where

$f(0)$ and $s(0)$ = corresponding values at $t = 0$

If new chains are formed, some may participate in load sharing and some may not. Hence, in a rubber maintained at a fixed extension, there will be two types of chain networks. One kind of network is in equilibrium with the stretched condition and the other type of network is in equilibrium with the unstretched condition. If unstretched and stretched lengths are denoted by l_u and l_x and the final length or set length by l_s, then we may write for two sets of lengths the following equations:

$$f_u = s_u \, kT[(l_s/l_u)^2 - (lu/ls)] \tag{3.15}$$

$$f_x = s_x \, kT[(l_x/l_u)^2 - (l_x/l_s)] \tag{3.16}$$

Where

f_u = stress per unit of cross-sectional area, which has its equilibrium at unstretched length

s_u = number of those chains

f_x and s_x = corresponding stress and the number of network chains in the stretched condition which are in equilibrium at the extended length

For equilibrium at permanent set:

$$f_u = -f_x \tag{3.17}$$

$$s_u \, l_x^2 / s_x \, l_u^2 = (l_x^3 - l_s^3)/(l_x^3 - l_u^3) \tag{3.18}$$

$$\% \text{ permanent set} = \frac{(l_s - l_u)}{(l_x - l_u)} \times 100 \tag{3.19}$$

$$\% \text{ permanent set} = \frac{(l_s/l_u - 1)}{(l_x/l_u - 1)} \times 100 \tag{3.20}$$

$$l_s/l_u = \left[\frac{(l_s/l_u)^3 - 1}{(s_u/s_x)((l_s/l_u)^2 + 1)} + 1 \right]^{1/3} \tag{3.21}$$

$$\% \text{ set} = \left\{ \left[\frac{C_1}{(s_u/s_x)C_2 + 1} + 1 \right]^{1/3} - 1 \right\} C_3 \tag{3.22}$$

The stress values given by the continuous stress relaxation curve are proportional to s_u and those yielded by the intermittent stress relaxation curve are proportional to $s_u + s_x$

If f/f_0 yielded by the continuous stress relaxation curve is designated as U and the difference in f/f_0 values of the intermittent and continuous stress relaxation curve are designated as X, then:

$$s_u/s_x = U/X \tag{3.23}$$

$$\% \text{ set} = \left\{ \left[\frac{C_1}{(U/X)C_2 + 1} + 1 \right]^{1/3} - 1 \right\} C_3 \tag{3.24}$$

$$C1 = (l_x/l_u)^3 - 1$$

$$C2 = (l_x/l_u)^2$$

$$C3 = 100/(l_x/l_u) - 1$$

3.7.1.2 Compressive Stress Relaxation

The properties of rubber viscoelasticity that are important with reference to gaskets and seals are compressive stress relaxation, as they give an idea about the sealing capacity of rubber seals. Seals between two surfaces function by exerting a force on the upper surface, thus sealing the space between two surfaces, for example, two plates. Compressive stress relaxation measures the force that the rubber seal exerts on the upper surface as a function of time.

Compression set arises from compressive stress relaxation. Based on molecular theory or two network theories of the permanent set for stress relaxation at elevated temperatures, it can be argued that the same mechanism holds for compressive stress relaxation and compression set [25, 26].

Compressive stress relaxation test is preferred for estimating the service life of gaskets and seals. This test under accelerated conditions allows behavior prediction at ambient temperatures using the Arrhenius model. An acceptable value under these accelerated conditions serves as a quality parameter for the seal.

3.7.2 Hysteresis and Strain Energy

It is known that the energy imposed on an ideal elastic material is completely recovered when the force is removed. If the material is viscoelastic, the viscous component, through internal molecular friction, retards elastic deformation and some energy is lost. This lost energy is dissipated in the form of heat generated in the elastomer. Although hysteresis is observed in unfilled vulcanizates, it increases in filled vulcanizates.

Hysteresis is related to energy loss in a loading and unloading cycle and is observed as a loop. At a lower initial extension (loading cycle) of about 100%, vulcanized gum natural rubber exhibits very low energy loss. In comparison, at higher extension ratios to more than 300%, hysteresis loss becomes high and is attributed to strain-induced natural rubber crystallization. (A certain level of orderly arrangement of isoprene units of NR happens during loading and the order is lost during unloading of the rubber sample.) In carbon black filled vulcanizates, the agglomeration and deagglomeration of carbon black particles in the rubber matrix can happen during the loading and unloading cycle; hence, the filler also takes part in the energy dissipation process.

If an elastic polymer strip is subjected to a tensile force, the weak intermolecular forces of the polymer network are overcome and the molecular network becomes extended. A potential energy, elastic energy or strain energy becomes stored in the network. In the elastic region of extension, the ratio of stress to strain is a constant called the elastic modulus or Young's modulus (E).

In the polymer strip of length L, the cross-sectional area A is extended to a length l by force F and a restoring force appears in the region of linear viscoelasticity. In this deformation in the linear elastic region, the ratio of stress to strain is the elastic modulus (E):

$$F = \frac{EAl}{L} \tag{3.25}$$

The work done (dw) for the small extension when the strip is extended by a small length dx is given by

$$dw = Fdx$$

and the total work done W in extending the strip to a length l is obtained by integrating this expression over the deformation range:

$$W = \int_{x=0}^{x=l} Fdx \tag{3.26}$$

For a small extension x, the elastic modulus is given by:

$$E = \text{stress/strain} = \frac{F/A}{x/L} \tag{3.27}$$

$$F = E\,A\,x/L \tag{3.28}$$

The work done then becomes:

$$W = \int_0^l \frac{EA}{L}x\,dx$$

$$W = \frac{EA}{L}\left[\frac{x^2}{2}\right]_0^l = \frac{1}{2}\frac{EA}{L}l^2 = \frac{1}{2}\left(\frac{EAl}{L}\right)l$$

$$\text{Work done} = \frac{1}{2}F \times l = \frac{1}{2}\text{load} \times \text{extension}$$

The energy generated by the work done during rubber deformation is stored as strain energy, which is quantified per unit volume of the material

Strain energy (S.E.) is in the form of work done:

$$\text{S.E.} = \frac{1}{2}\text{load} \times \text{extension} \tag{3.29}$$

Based on the volume (V) of the strip:

$$\text{S.E./volume} = \frac{1}{2}\frac{F}{A}\frac{l}{L} \tag{3.30}$$

The applied force generates strain energy, also known as elastic energy or elastic potential energy, which is stored in the polymer strip. Upon removal of the force, this energy is released as kinetic energy. The strain energy (S.E.) per unit volume in the polymer strip is:

$$\text{S.E. per unit volume} = \frac{1}{2} \text{stress} \times \text{strain} \qquad (3.31)$$

In terms of elastic modulus E:

$$\text{S.E. during extension per unit volume} = 1/2E \; e^2 \qquad (3.32)$$

The strain energy stored in the elastic deformation region is called resilient energy and this strain energy is represented graphically by the area under the load–deformation curve during retraction (Figure 3.2 b). The area under the curve as the rubber is stretched represents the work done on the rubber (Figure 3.2 a).

In a stress–strain graph, the area under the loading curve corresponds to the work done on the material – essentially the elastic potential energy stored. During unloading, the area under the unloading curve represents the energy released by the stored potential energy. The area enclosed between these two curves (the hysteresis loop) quantifies the energy dissipated during the cycle (Figure 3.2 c).

When rubber is subjected to cyclic deformation, energy losses occur, due to internal friction that arises from factors such as slippage of rubber chains through weak intermolecular interactions, weak interactions involving rubber molecules and fillers used in rubber, and movement of side chains that do not take part in the load bearing process. Hysteresis is determined as heat build-up by a forced vibration method or as loss tangent by dynamic mechanical analysis of rubber involving a sinusoidal deformation.

Heat build-up can be measured with a Goodrich flexometer that conforms to ASTM D 623 (2007). The cylindrical samples (2.5 cm in height and 1.9 cm in diameter) are tested under compression at 1800 rpm with a dynamic stroke of 4.45 mm and a static load of 1 MPa. For the blowout test, the static load is 2 MPa. The temperature rise or compression set can be measured. The blowout time for the rubber specimen can also be measured to evaluate hysteresis.

There are other items of testing equipment that measure the fatigue life of rubber, such as crack initiation, crack growth and complete failure following repeated flexing. The De Mattia Flex Tester, Ross Flex Tester and Fatigue Failure Tester are used extensively for fatigue tests.

The force required for extending and returning to an undeformed condition is different, as shown in Figure 3.2.

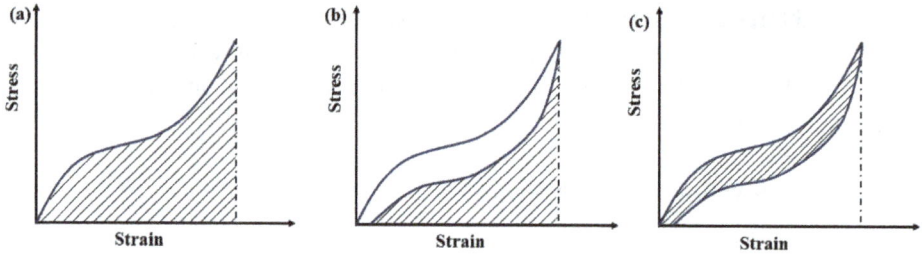

Figure 3.2 Rubber specimen subjected to cyclic deformation: (a) stress–strain graph for extension showing energy absorbed, (b) stress–strain graph for retraction showing the energy stored, (c) stress–strain graph showing energy dissipated in one cycle

3.7.3 Resilience

The energy stored during deformation of rubber by a force may not be recovered after the force is removed. The ratio of energy released upon removal of the deforming force to the energy stored during deformation by the force is called resilience and is expressed as a percentage.

One of the test methods is the impact test, which involves rebounding in different forms. A Dunlop tripsometer utilizes a rotating disk and hemispherical striker that strikes a rubber specimen and measures rebound height. The resilience is expressed as rebound resilience.

The Lupke pendulum involves striking a rubber specimen with a swing pendulum. The resilience is expressed as rebound resilience and is the ratio between the returned and the applied energy of a moving mass which impacts a test piece [27].

In the Bayshore resilience method, a weighted ball is dropped from a specified height onto a rubber sample and the rebound height is measured. Resilience is calculated as the ratio of the rebound height to the initial drop height.

Yerzley Mechanical Oscillograph can also be used to determine the resilience as per ASTM D 945. A cylindrical rubber specimen is mounted between the loading foot and anvil of a Yerzley Mechanical Oscillograph, which operates using a beam pivoted on a knife-edge fulcrum. When a calibrated weight is suddenly applied to the free end of the beam, the rubber deforms and initiates damped oscillations. These oscillations are recorded as a damped sinusoidal trace on a rotating drum or paper. The difference in amplitude between successive peaks of the trace is used to calculate the logarithmic decrement, from which the resilience of the rubber is determined.

3.7.4 Cyclic Deformation of Rubber: Mullins Effect and Payne Effect in Carbon-Black-Filled Systems

When a rubber specimen is subjected to a cycle of loading and unloading to a certain degree of deformation, followed by the repetition of this cycle of loading and unloading to the same deformation, it will be observed that in the second cycle of deformation, the load required for the same deformation is lower than the load required in the first cycle of loading. In the initial loading, a vulcanizate appears stiff. When this vulcanizate is unloaded and reloaded, the stress–strain follows a lower path and, after many repeated cycles, the stress–strain curves stabilize. This decrease in load in the cyclic deformation of rubber is called the strain-softening effect or Mullins effect [28, 29]. This stress decay in cyclic deformation becomes more prominent in carbon-black-filled vulcanizates. In carbon-black-filled vulcanizates, it can be assumed that there is a soft region constituting vulcanized rubber molecules and a hard phase consisting of vulcanized rubber molecules in weak interaction with carbon black. So, in the second loading cycle, the soft region in the matrix increases and the load required for the same deformation becomes less. There is also slippage of polymer chains on the surface of the filler. Mullins and Tobin reported that, as the filler content increases, the strain amplification factor (X) also increases compared with pure rubber vulcanizates [30]. The presence of filler amplifies the strain in a rubber. The locally applied strain will be larger than the actual strain, because rigid filler does not undergo strain. This is called strain amplification. Many types of filler rubber interaction exist in a rubber vulcanizate, due to the high surface area of fillers, structure and certain functional groups present on the filler surface. The Mullins effect can be caused by changes in these filler rubber interactions, destruction of the filler aggregate network, molecular slippage and disentanglement. Although the softening is present in elastomers with and without fillers, the effect is smaller in intensity in unfilled ones. In gum vulcanizates, the Mullins effect may originate in disentanglements of polymer chains during deformation [31].

The Mullins effect exhibits time-dependent behavior. As the frequency of deformation increases, stress softening increases. The increased stress softening with strain rate may be due to the entanglements locked in a network and chain slippage (molecular friction) in the bulk rubber and the filler interface. Increased contribution of scission at a higher strain rate can also lead to energy dissipation by rubber damage. The elastic network can become damaged during deformation by breakage of the main rubber, scission of cross-links or rupture at the rubber filler interface. This is a phenomenon related mainly to filler–rubber interactions.

The Payne effect shows that the elastic modulus depends on the deformation and decreases at a higher amplitude of deformation. Filler particles such as silica can have filler–filler interactions (due to physical bonds, such as Van der Waals) that can rupture as the amplitude of deformation increases. The destruction of the filler network is responsible for the Payne effect in silica or carbon-black-filled rubber [32]. The Payne effect depends on the filler content of the material and vanishes for unfilled rubber vulcanizate. A lower Payne effect indicates better filler dispersion, as filler–filler interaction reduces when filler dispersion is good.

3.7.5 Phase Lag

Relaxation occurs during stress relaxation and creep when stress or strain is kept constant. It also happens under sinusoidal deformation, where it is quantified by the lag between stress and strain. Greater phase lag indicates more energy loss or hysteresis.

3.7.5.1 Dynamic Mechanical Properties of Rubber

Sinusoidal deformation is used to study the behavior of rubber under dynamic conditions, which is based on observing the deformation of rubber under the action of a force varying sinusoidally with time. A dynamic mechanical analyzer is used to study the deformation of a rubber sample of a suitable dimension. The sinusoidal deformation is used to obtain modulus plots as a function of temperature, frequency of deformation or time of deformation. Deformation occurs in linear viscoelastic regions where oscillatory strain varies linearly with stress.

The angular frequency ω of deformation describes the sinusoidal deformation (angular frequency is related to the period T which is the time taken for one oscillation, frequency $f = 1/T$ and angular frequency $\omega = 2\pi/T$. The unit of ω is rad/s). In viscoelastic materials, the response is out of phase with the stress applied by an angle δ, called the phase angle. This phase angle occurs because more time is required for molecular chains to return to their equilibrium position or for the relaxation and motion of the rubber chains. A viscoelastic material can be studied by subjecting a rubber specimen to sinusoidal deformation $\epsilon(t)$ of an angular frequency ω given by:

$$\epsilon(t) = \epsilon_0 \sin(\omega t) \tag{3.33}$$

Where

ϵ_0 = maximum strain amplitude

ω = angular frequency (2π times the frequency in Hertz)

$\epsilon(t)$ = strain at a time t

The shear response $\sigma(t)$ is also sinusoidal and is out of phase with strain. This phase shift between stress and strain exists because a part of the energy input is not recovered at the end of the cycle. Some energy input is stored and recovered in each cycle and some energy is lost through friction and dissipated as heat. This energy loss can have various reasons related to the viscous flow of rubber and the breaking of the weak rubber filler interactions related to the morphology and structure of fillers. This shear response can be mathematically expressed as shown below (0 is the maximum shear response amplitude):

$$\sigma(t) = \sigma_0 \sin(\omega t + \delta) \tag{3.34}$$

Where

σ_0 = shear response at maximum strain

δ = phase angle

This phase angle is illustrated in Figure 3.3.

$$\sigma(t) = \sigma_0 \sin(\omega t + \delta) = (\sigma_0 \cos \delta) \sin \omega t + (\sigma_0 \sin \delta) \cos \omega t \tag{3.35}$$

σ_0 is the shear response at maximum strain and δ is the phase angle. The shear stress signal (t) thus has two different contributions, because it is split into a sine function and a cosine function (which has a phase difference of 90°) as one is in phase with the strain $\sigma_0 \cos \delta$ and the other is 90° out of phase with the strain $\sigma_0 \sin \delta$. The stress realized in the sinusoidal deformation of rubber samples in a dynamic mechanical analyzer shows that the total stress has two components *viz.*, elastic stress and viscous stress, as shown in Figure 3.3.

These two contributions can be written in terms of the dynamic modulus. The modulus corresponding to the in-phase component is designated E' and the one out of phase with strain is designated E''. E' is called the elastic modulus or storage modulus while E'' is the loss modulus or viscous modulus, as given in Equation 3.36:

$$\frac{\sigma_0}{\epsilon_0} \cos \delta = E' \tag{3.36}$$

And

$$\frac{\sigma_0}{\epsilon_0} \sin \delta = E''$$

Applying Equation 3.36 to Equation 3.35

$$\sigma(t) = \epsilon_0 [(E') \sin \omega t + (E'') \cos \omega t] \tag{3.37}$$

The vector sum of these two components yields the complex modulus. The dynamic properties can be decomposed into storage and loss modulus E'' or complex modulus E^* and phase angle δ. The vector diagram for the viscous and elastic components is shown in Figure 3.3.

The out-of-phase modulus is considered imaginary, so that the complex modulus E^* is given by:

$$E^* = E' + i E'' \tag{3.38}$$

The two components of the complex modulus represent the real and the imaginary part of the shear modulus E^* when referred to as a complex modulus. The real part corresponds to the ability to store energy, while the imaginary part corresponds to the ability to dissipate energy.

Consider the modulus in terms of vectors.

In terms of complex modulus E^*, the phase angle δ is:

$$E' = E^* \cos \delta \tag{3.39}$$

$$E'' = E^* \cos \delta \tag{3.40}$$

Phase angle is defined by:

$$\frac{E''}{E'} = \tan \delta = \frac{\text{Viscous modulus}}{\text{Elastic modulus}} \tag{3.41}$$

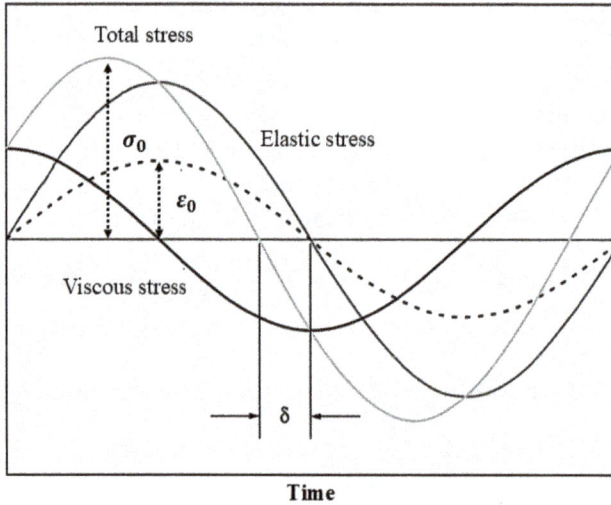

Figure 3.3 Stress behavior in rubber under sinusoidal deformation

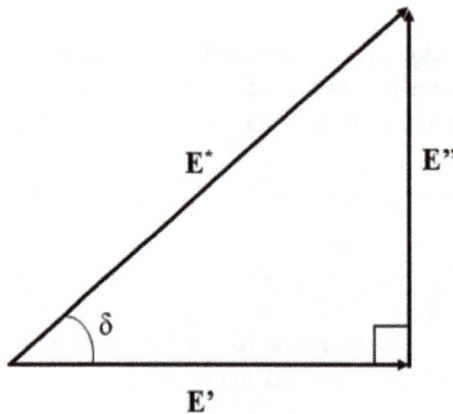

Figure 3.4 Vector representation of complex modulus, elastic modulus and viscous modulus

3.8 Viscoelasticity – Mathematical Models

Two mechanical models are used to characterize viscoelastic behavior, *viz.* the Maxwell model and the Voigt model, represented in Figure 3.5. These models are used to predict the different viscoelastic properties. Creep behavior and the relaxation process under sinusoidal deformation over a wide frequency range can be modeld with the Voight model. In contrast, stress relaxation can be modeled with the Maxwell model.

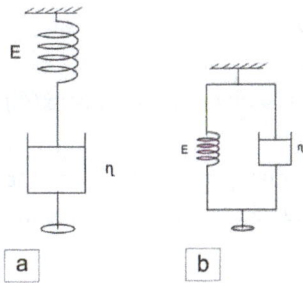

Figure 3.5 Schematic representation of viscoelastic models: a) Maxwell model, b) Voigt model

3.8.1 Maxwell Model

The Maxwell model consists of a spring (the elastic component) and a dashpot (the viscous component) connected in series. With this type of arrangement, the stress is assumed to be uniformly distributed and the stress realized by the viscous component is the same stress realized by the elastic component. The stress developed in the elastic component, σ(elastic), is equal to the stress developed in the viscous component σ(viscous).

When an external force is applied, both elastic and viscous components respond separately. The total strain will be the sum of strain experienced by elastic and viscous components. Under a constant applied strain, the stress increases instantly to a maximum value determined by the elastic modulus of the spring and then relaxes exponentially.

An ideal elastic material can be represented in terms of stress σe and strain ε as follows.

$$\sigma e = E\epsilon$$

Where

E = a constant called Young's modulus or elastic modulus and is related to the material's stiffness

An ideal viscous component shear stress (σv) is related to the strain rate, $\dot{\epsilon}$, and mathematically expressed as follows:

$$\sigma v = \eta \dot{\epsilon}$$

Where

η = viscosity of the viscous component

The total stress is:

$$\sigma = E\epsilon = \eta \dot{\epsilon} \tag{3.42}$$

(In the Maxwell model, total stress $\sigma = \sigma$ of elastic component $= \sigma$ of viscous component)

Total strain in the system is the sum of the strain in the viscous ($\dot{\epsilon}_v$) and elastic ($\dot{\epsilon}_e$) components.

Since the viscous component has a derivative term, a similar physical term for the elastic component is needed in order to obtain a constitutive equation to describe the Maxwell model:

$$\sigma = E\epsilon_e \tag{3.43}$$

So:

$$\dot{\sigma} = E\dot{\epsilon}_e$$

$$\dot{\epsilon}_e = \frac{1}{E}\dot{\sigma} \tag{3.44}$$

For the viscous component, the strain rate is mathematically expressed as follows.

$$\dot{\epsilon} = \sigma\frac{1}{\eta} \tag{3.45}$$

Based on the strain, the constitutive equation for the Maxwell model is:

$$\dot{\epsilon} = \frac{1}{E}\dot{\sigma} + \sigma\frac{1}{\eta} \tag{3.46}$$

Multiplication by E and rearrangement yields the Maxwell model in the form of a differential equation.

$$E\dot{\epsilon} = \dot{\sigma} + \sigma(E/\eta) \tag{3.47}$$

$$\dot{\sigma} + (E/\eta)\sigma - \dot{\epsilon}E = 0 \tag{3.48}$$

This model can be used to explain stress relaxation, which is the reduction in stress when strain is constant ($\dot{\varepsilon} = 0$). Hence, Equation 3.48 can be written as:

$$d\sigma/dt = -(E/\eta)\sigma \tag{3.49}$$

Rearrangement yields:

$$d\sigma/\sigma = -E\,dt/\eta \tag{3.50}$$

This equation can be integrated to obtain the stress decay within limits σ_0 and σ_t, and time within limits of 0 to t.

$$\int_{\sigma 0}^{\sigma t} d\sigma/\sigma = \int_{0}^{t} -(E/\eta)dt \tag{3.51}$$

Where

$t = 0$

$\sigma = \sigma_0$

$$\ln \sigma_t/\sigma_0 = -(E/\eta)t \tag{3.52}$$

Or, in exponential form:

$$\sigma(t) = \sigma_0 e^{-t/\tau} \tag{3.53}$$

Where

τ = relaxation time, which is given by η/E

Given that $\tau = t$,

$$\sigma(t) = \sigma_0 e^{-1} \tag{3.54}$$

The relaxation time τ is physically the time needed for the stress to fall to $1/e$ or about 37% of its initial value, as seen in Figure 3.6.

The stress relaxation modulus E_R may be obtained from this relation directly:

$$E_R = \sigma_t/\varepsilon_0 = (\sigma_0/\varepsilon_0)e^{-t/\tau} \tag{3.55}$$

The relaxation time τ is very sensitive to temperature change and factors related to chain flexibility and is roughly the inverse of the rate of molecular motion. Above T_g, relaxation is fast with a short τ and, below T_g, it is very slow, with a long τ, as illustrated in Figure 3.6.

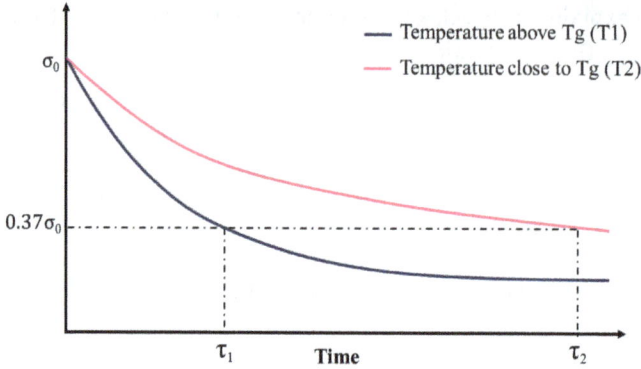

Figure 3.6 Stress relaxation for the Maxwell model at temperatures close to T_g and above T_g

A term similar to relaxation time is the relaxation modulus:

$$E_{rel} = \frac{\sigma(t)}{\varepsilon_0} \times e^{-t/\tau} \tag{3.56}$$

Stress relaxation modulus is a time-dependent property shown by viscoelastic materials and varies with temperature.

If the same model is used to describe creep, where stress is a constant ($d\sigma/dt = 0$), Equation 3.46 can be written as

$$d\varepsilon = (\sigma/\eta)dt \tag{3.57}$$

The strain increase from an initial ε_0 to ε in a time t.

The strain in the sample after time t can be obtained by integration as shown below

$$\int_{\varepsilon 0}^{\varepsilon} d\varepsilon = \int_{0}^{t} \sigma/\eta dt$$

On integration we get:

$$\varepsilon = \varepsilon_0 + \sigma t/\eta \tag{3.58}$$

The equation is linear, but the actual creep behavior in rubber is exponential. So, this model cannot explain the creep behavior of rubber.

3.8.2 Voight Model

In the Voigt model, the viscous and elastic components are connected in parallel. In this case, it is assumed that there is a uniform distribution of strain, so the spring and the dashpot deformations are equal. The force experienced is different for the viscous and elastic components. The total force is the sum of force experienced in the viscous and elastic components.

This model is useful for explaining the creep observed in viscoelastic materials.

The total stress is given by:

$$\sigma = E\varepsilon + \eta d\varepsilon/dt \tag{3.59}$$

Division by E yields:

$$\sigma/E = \varepsilon + \eta(d\varepsilon/dt)/E \tag{3.60}$$

$$(\sigma/E) - \varepsilon = (\eta/E)(d\varepsilon/dt) \tag{3.61}$$

$$\frac{d\varepsilon}{(\sigma/E) - \varepsilon} = (E/\eta)dt$$

This differential equation has a solution by integration as given below:

$$-\ln((\sigma/E) - \varepsilon) + C = (E/\eta)t \tag{3.62}$$

In a creep experiment, at $t = 0$, $\varepsilon = 0$, then $C = \ln(\sigma/E)$

$$-(E/\eta)t = \ln((\sigma/E) - \varepsilon) - \ln(\sigma/E) \tag{3.63}$$

$$-(E/\eta)t = ln(1 - \varepsilon/(\sigma/E)) \tag{3.64}$$

Expressed in exponential form:

$$e^{-(E/\eta)t} = 1 - \varepsilon E/\sigma \tag{3.65}$$

$$\varepsilon E/\sigma = 1 - e^{-(E/\eta)t}$$

In terms of strain, at a time t:

$$\varepsilon(t) = (\sigma/E)[1 - e^{-(t/(E/\eta)}] \tag{3.66}$$

E/η is the retardation time denoted as λ_r and is the rate at which retarded elastic deformation occurs

$$\varepsilon(t) = (\sigma/E)[1 - e^{-t/\lambda_r}]$$

Assuming a $\lambda_r = t$, the equation becomes:

$$\varepsilon(t) = (\sigma/E)[1 - e^{-1}] \tag{3.67}$$

The retardation time is the time required to deform to a value of $(1 - (1/e)$ or about 63% of its total deformation. In a viscoelastic model, retardation time reached during the application of a constant stress during time t_0 to t_1 is illustrated in Figure 3.7.

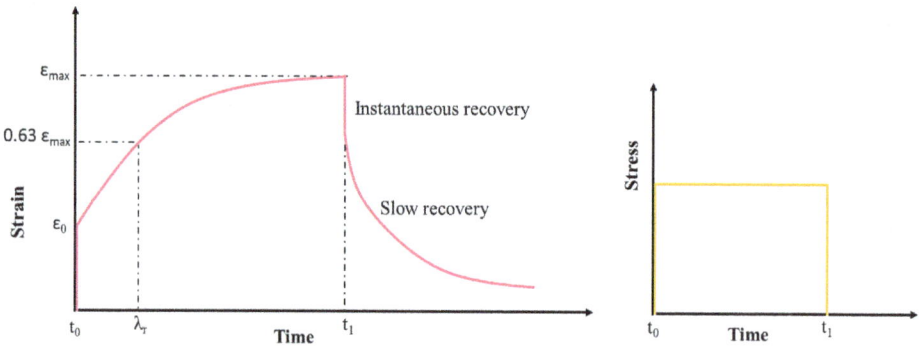

Figure 3.7 Behavior of a viscoelastic material subjected to a creep test based on the Voight model and showing the retardation time

As per this model, if a constant initial stress is applied to a specimen, it will continue to deform for as long as the stress is applied and the specimen will recover to some extent after the removal of the stress.

These are only mathematical models based on Hooke's law and Newtonian flow, and they may not interpret the behavior at the molecular level in practical observations, possibly due to the specific micro-and macrostructure of rubber and the different additives added in processing.

3.9 Influence of Temperature on Viscoelastic Parameters

In rubber molecules, the ability to rotate the carbon–carbon bonds can give rise to localized motion of chain segments, leading to entanglements in the rubber chain network. The entanglements allow the motion of chain segments as the temperature increases. With decrease in temperature, localized movement of rubber chains becomes restricted, which can be attributed to the lowering of free volume. As the temperature lowers, these entanglements behave as cross-links.

This relaxation process is reflected in the elastic modulus values of rubber. As the temperature lowers to the T_g range, the ability to undergo this conformational change decreases and rotational motion appears to be frozen. The lower the temperature and the closer to the T_g, the higher the value of the elastic modulus. The greater the difference between T_g and a particular temperature above T_g, the lower the modulus and the greater the chain flexibility. Hysteresis close to the glassy region is high, as chain flexibility is very low.

At low temperatures, the modulus of rubber is high, making it brittle and rigid. At this temperature, rubber as such has no rubber-like property. At higher temperatures, the modulus of rubber is low, making it flexible and elastic. At very low temperatures, rubber undergoes a second-order transition to glass. As a viscoelastic material, the variation in vulcanized rubber's stress and response can be distinguished in three regions. They are: 1) glassy region where the elastic modulus is high, 2) glass transition region where the modulus changes by several decades, and 3) rubbery region where the elastic modulus is low and rubber undergoes high extension.

3.9.1 WLF Equation (Williams–Landel–Ferry Equation)

The influence of time and temperature on certain viscoelastic parameters of polymers, especially in the glass transition region and above, can be studied by the WLF equation.

In general, an Arrhenius equation can describe the temperature dependence of the viscosity of a viscous fluid, such as honey.

$$\eta = Ae^{E/RT} \tag{3.68}$$

Where

T = absolute temperature

R = universal gas constant

A = material constant

E = activation energy

Knowledge of the two material-dependent parameters, E and A, allows the viscosity at any temperature to be predicted. The two parameters are usually determined from a plot of log η against $1/T$, which yields a straight line:

$$\ln \eta = \log A + E/RT \tag{3.69}$$

This equation is not valid for most polymers, especially those with high molecular weights, as the log plots are not linear over a wide temperature range of more than about 50 °C. For high-molecular polymers, viscosity varies as a function of free volume rather than activation energy. It is assumed that free volume changes linearly with temperature (Figure 3.8).

Viscosity is an exponential function of the reciprocal of the fractional free volume fr, which is the ratio of free volume (Vf) to total volume (V) [33, 34].

At a temperature T,

$$fr = Vf(T)/V \qquad\qquad (3.70)$$

$$\eta = AeB/(Vf/V) \qquad\qquad (3.71)$$

Where

A and B = constants

The empty space between polymer molecules is called free volume (Vf).

The volume occupied by polymer molecules is called occupied volume (Vo). So, the volume of a polymer, called the specific volume, denoted by V, can be mathematically expressed as:

$$V = Vo + Vf \qquad\qquad (3.72)$$

Where

V = volume occupied by a polymer

Temperature affects free volume. As the temperature increases, free volume increases and as temperature decreases, free volume decreases. At a particular low temperature, the free space between molecules becomes so small that there will be practically no significant molecular movements (perhaps some vibrations in their own space) and this temperature corresponds to T_g.

Above T_g for an amorphous polymer, the fractional free volume is given by:

$$fr = fr_g + (T - T_g)\alpha \qquad\qquad (3.73)$$

Where

α = expansion coefficient when T greater than or equal to T_g

fr_g = fractional free volume at T_g

$$\ln \eta(T) = \ln A + B/fr \qquad\qquad (3.74)$$

In this equation, when $T = T_g$:

$$\ln \eta(g) = \ln A + B/fr_g \qquad\qquad (3.75)$$

$$\ln \frac{\eta(T)}{\eta(T_g)} = B\left[\frac{1}{fr} - \frac{1}{fr_g}\right] \qquad\qquad (3.76)$$

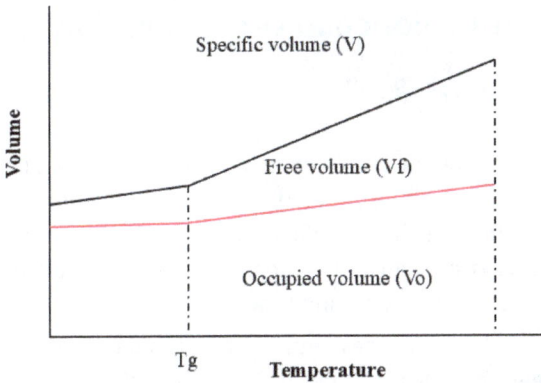

Figure 3.8 Schematic representation of change in free (specific volume) volume of a polymeric material with temperature

As the temperature of the polymer is lowered, the free volume will eventually reduce and there will not be enough free volume to allow the molecular rotation to take place. The temperature at which this happens corresponds to T_g as, below this temperature, the polymer glass is effectively frozen.

Williams, Landel and Ferry have suggested that the viscosity η at a temperature T may be related to the viscosity η_g at the glass transition temperature: The ratio of viscosities is called the shift factor.

$$\ln a_T = \ln \frac{\eta(T)}{\eta(T_g)} = B \left[\frac{1}{fr_g + \alpha(T - T_g)} - \frac{1}{fr_g} \right] \tag{3.77}$$

Simplifying and substituting:

$$\frac{B}{2.303 fr_g} = C1$$

And

$$fr_g / \alpha = C2$$

$$\log a_T = \log \frac{\eta(T)}{\eta(T_g)} = -\frac{C1(T - T_g)}{C2 + (T - T_g)} \tag{3.78}$$

This equation implies that properties, such as viscosity, when expressed on a logarithmic scale at two temperatures, one at a reference temperature T and one at glass transition temperature T_g, can be related by a factor a_T, called the shift factor. So, it can be assumed that different polymers can have the same viscosity at T_g. The viscosity of a polymer at any temperature above T_g can be determined if the value of viscosity at T_g and the two WLF constants $C1$ and $C2$ are known.

3.10 Time–Temperature Equivalence in the Relaxation Process

The stress relaxation process is time-dependent and is primarily due to the viscous component of a viscoelastic polymer. It can be considered to be due to the flow of polymer chains that lead to a decrease in stress in the deformed condition. As per the Maxwell model, the relaxation process at a particular temperature decreases exponentially with time and becomes stabilized after a certain time.

When a material is subjected to external forces, it responds with internal changes at the atomic and molecular levels. These changes include fluctuations in the electron cloud, alterations in bond length or rotation, and for elastomers, segmental motion, side-chain motion, or conformational changes of the entire network. The type of relaxation mode in a viscoelastic material depends on the temperature and the rate at which the load is applied.

For natural rubber, the relaxation rate can depend on factors such as the gel content, branching and molecular weight. The relaxation process that occurs in the glassy region, glass-rubber transition region and rubbery region are different for rubbers. In the rubbery state, the relaxation can be segmental relaxation or α relaxation. In glassy regions, the rubber chains are in a frozen state and segmental relaxation is not possible. The stress relaxation modulus increases as temperature increases.

Viscoelastic responses are measured using stress, strain, or oscillatory methods at various deformation frequencies. While conventional equipment can determine properties like stress relaxation or dynamic modulus, it is limited to a narrow range of time or frequency.

For products like hoses, gaskets, and seals, it's crucial to understand viscoelastic parameters over extended periods or a wider range of deformation frequencies. The Time-Temperature Equivalence (TTE) principle allows for this by assuming that the effects of changing time or frequency are equivalent to changing the temperature. This allows us to predict long-term material behavior from short-term tests. Based on this equivalence we can construct a master curve for a particular parameter at extended time scale or frequency scale using the shift factor. This shift factor can be calculated using WLF equation .

From the WLF equation, it is seen that the viscosity of a polymer at temperatures T above T_g is primarily dependent on the free volume of the polymer at that temperature T. The free volume of a polymer above T_g increases linearly with temperature. The main factor that determines viscosity at a particular temperature T above T_g is the difference in temperature between them $(T - T_g)$ and not the activation energy E, which is an energy barrier as per the Arrhenius equation.

A superimposed relaxation modulus curve at a particular temperature can be obtained from stress relaxation data collected in a short duration of time at different temperatures.

Figure 3.9 (a) shows isothermal stress relaxation modulus curves as a function of time at five different temperatures (T_1 to T_3) while Figure 3.9 (b) illustrates how the curves are shifted horizontally using a shift factor to obtain the change in relaxation modulus over a wider time range at a suitable temperature (here: temperature T_2).

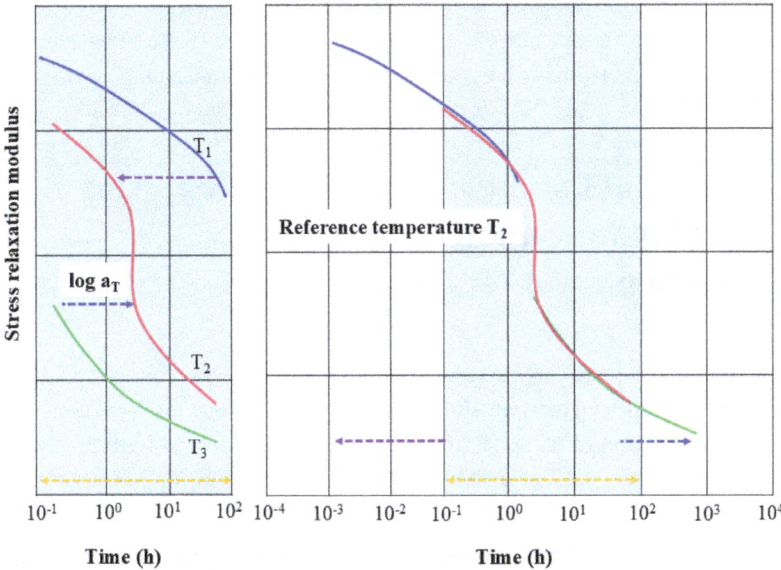

Figure 3.9 (a) Isothermal curves of loss modulus as a function of frequency at three temperatures T_1, T_2 and T_3, (b) Curve obtained by shifting isothermal curves T_1 and T_3 to a reference temperature T_2

The shift factor shifts a curve at a particular temperature T to become part of the curve at another temperature T_2. How much the curve at temperature T_1 is shifted to become part of the curve at T_2 is known as the temperature shift factor. In terms of the relaxation process, the shift factor is defined as:

$$a_T = t\tau_1/t\tau_2 \tag{3.79}$$

Where

$t\tau_1$ = time for a relaxation process that occurred in a time range at temperature T_1

$t\tau_2$ = time for a similar relaxation process that occurred in a time range at temperature T_2

The shift factor a_T refers to shifting the time on the time axis to obtain the same relaxation modulus at two temperatures.

Use of the shift factor widens the time axis, covering a longer time duration. This wider range of time can be termed the reduced time for a temperature T_2, which is considered as the reference temperature here. The time–temperature superposition technique shifts the isothermal curves horizontally to the left or right along the time axis until they overlap the next curve. The shift factors in terms of time on the X-axis are different for temperatures (T_1 to T_2 and T_3 to T_2). The WLF equation is used for amorphous polymers to understand the relaxation process related to parameters such as viscosity and stress relaxation at different temperatures above T_g up to $T_g + 100$ °C. The viscoelastic parameter generally used for rubbery polymers is the relaxation modulus.

3.11 Dependence of Dynamic Mechanical Properties on Time of Deformation or Frequency of Deformation and Temperature

Stress relaxation and creep tests can be conducted over long periods (hours or days), while oscillatory tests are performed on shorter timescales. At very low frequencies, materials have sufficient time to respond to the deformation, unlike at high frequencies. Oscillatory deformations allow for the separate measurement of elastic and viscous responses.

The parameters of dynamic mechanical analysis are generally obtained at a fixed deformation frequency as a function of temperature ranging from below T_g to a temperature higher than ambient. Suppose the same parameters are determined at higher deformation frequencies; the curves obtained will be the same, but will be shifted to a higher temperature (to the right on the temperature scale). At high deformation rates or low temperatures near the T_g, rubber exhibits a glass-like response. In this state, the polymer chain segments are "frozen" and lack the thermal energy needed for large-scale movement. At higher temperatures, however, the chains can undergo conformational changes at a very high rate. Under such conditions, particularly at high frequencies or low temperatures, the material exhibits a greater viscous response, resulting in increased energy loss.

Under low-frequency deformation, segmental movements or vibrations become easier and so does the relaxation process, but at higher deformation frequency, a proper segmental relaxation process leading to an equilibrium condition cannot happen and rubber shows transitions to a glassy state.

This time–temperature equivalence technique is significant, because the deformation frequency experienced by rubber products like tires and vibration isolators can be very high and it may not be possible to determine the actual values of dynamic properties at these high frequencies using conventional testing equipment. The equipment generally used, such as a dynamic mechanical analyzer, is not designed for conducting tests under very high-frequency deformation.

It has been observed that, under high-frequency cyclic deformation, the modulus is high and that, under low-frequency deformation, the modulus is low at a particular temperature. Or, under a high frequency of deformation, rubber tends toward a glassy state with more hysteresis; under low-frequency deformation, rubber retains the rubbery nature with low hysteresis. The modulus obtained at a particular frequency and temperature can be the same as that obtained at another frequency and temperature.

When tests are performed at low frequencies to measure properties such as the loss tangent at various temperatures, it becomes possible to extrapolate data over a wide frequency range if the shift factor corresponding to a specific frequency variation is known. By leveraging the principle of time–temperature equivalence (TTE), the loss tangent value at very high frequencies can be estimated through low-frequency testing.

The mobility of rubber chains plays a crucial role in rubber being rubbery or glassy in nature. In sinusoidal deformation, a slow cycle of deformation (low-frequency deformation), the polymer chains respond to the applied stress without delay and energy loss. As the cycle of deformation becomes fast (frequency higher), the rubber chains lose their ability to respond to applied force without delay and energy loss happens in a cycle of deformation. This is represented schematically in Figure 3.10.

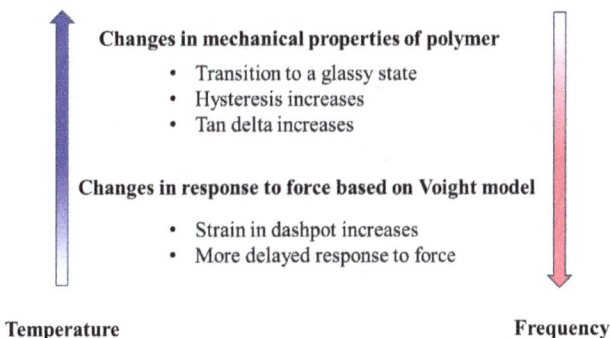

Changes in mechanical properties of polymer

- Transition to a glassy state
- Hysteresis increases
- Tan delta increases

Changes in response to force based on Voight model

- Strain in dashpot increases
- More delayed response to force

Temperature   Frequency

Figure 3.10 Variation of viscoelastic properties with changes in deformation frequency and temperature

When a high-frequency deformation is imposed on rubber, energy absorption is a maximum and rubber behaves as if in a glass transition zone [35]. Behavior close to that of the glassy state is observed at ambient temperatures when rubber is subjected to cyclic deformation with a high frequency above 10^6 Hz and wet skidding can be associated with this deformation frequency [36]. At these frequencies, the polymer molecules do not have enough time to respond to external deforming forces. Under a medium range of deformation frequency, the rubber becomes leathery and, under low deformation frequency, the behavior is rubbery.

Rubber is more elastic under low-frequency deformation and more viscous under high-frequency deformation.

The practical significance of TTE is that various tire performance parameters, especially road grip when the brakes are applied and where the frequency of tire deformation could be in the MHz range, can be predicted from data determined at a low-frequency range of close to 1 Hz. A tire needs different parameters for good performance, but can be condensed to magic triangle properties: rolling resistance relative to fuel efficiency, traction relative to wet and dry grip, and abrasion resistance relative to mileage and durability.

The uniqueness of rubber viscoelasticity is that, at low deformation frequency, rubber is more elastic, while at high deformation frequency, rubber exhibits high hysteresis. This is useful, because rubber needs to be elastic and viscous at the same time, depending on input frequencies in products such as tires and vibration isolators.

3.11.1 Time–Temperature Equivalence: Significance in the Prediction of Properties

Rolling resistance

Rolling resistance is the resistance force that develops when the tire rolls. The force that opposes the rolling of a tire attached to the wheel quantifies the retarding effect of the road surface at the tread and road interface. In terms of force, the force needed to keep a free-rolling tire at a constant speed can be called the rolling resistance, though it is known that equilibrium needs to be maintained by the different forces that come into play during the rolling of a tire. In terms of energy loss, rolling resistance is the energy lost when the tire rolls a certain distance. The elastic properties of the tread (depending on construction), friction at the tread and road contact interface, and the type of road, whether hard or soft, are the three important factors that govern the rolling of the tire.

When a tire rolls on a surface, the surface exerts a force on the tire perpendicular to the surface. This is the normal force (F_Z). Rolling resistance as a force (F_R) is expressed as:

$$C_R = F_R/F_Z \tag{3.80}$$

Where

C_R = coefficient of rolling resistance

The energy lost during rolling of the vehicle tires is due to various factors, such as deformation of tire tread, deflection of different tire components, abrasion loss while rolling, difference in car speed and tire rotation speed (called tire slip), air drag and energy loss due to irregularities on road surface. When a tire rolls, a rise in temperature can also affect the rolling resistance. A rise in temperature from ambient by about 30 °C reduces it. As the air pressure inside the tire increases, the rubber at the contact patch becomes stiffer, reducing its deformation and hence reducing rolling resistance. The hysteresis also reduces slightly. High elasticity and low hysteresis are required for low rolling resistance.

In addition to good rolling ability, a tire should have good grip or adhesion to dry and wet road surfaces. Therefore, grip is used synonymously with traction, which is the ability of a tire to hold the road without skidding. When a tire rolls, a contact patch is formed and the friction at the contact patch allows the vehicle to accelerate, brake and avoid skidding and cornering.

Grip

A tire can have good grip if the stopping distance is short. The kinetic energy of the moving vehicle is reduced to zero when the brake is applied; it is dissipated mainly by the tires of a vehicle due to its viscoelastic nature. In fact, other factors such as aerodynamic drag, inertia force and frictional forces offered by the road surfaces resist the rolling of the tire and other vehicle engine-related forces that influence tire grip or traction or skid resistance. The hysteresis generated during deformation of the tire, the abrasion of the tread and the adhesive nature of rubber all combine to provide good road grip. High hysteresis prevents the tire from sliding.

When the tire rolls and touches the ground, a portion of the tread becomes compressed (called a contact patch), due to deformation, but it can regain its shape immediately as the tire rolls away from the road surface, because the deformation of rubber has an elastic component that has a response which is instantaneous and fully recoverable.

The deformation of rubber also involves a viscous component which is not instantaneous (time-dependent) and not fully recoverable. The shape of the compressed tire tread may not be fully recovered in one rotation, due to viscous components. The energy not recovered is converted into heat in the tire.

Two parameters related to grip are adhesion and friction force at the contact patch.

Grip is commonly described in terms of the coefficient of friction (C_f), which is defined as the ratio of frictional force (F_f) and normal force (F_N):

$$C_f = F_f / F_N \tag{3.81}$$

Where

F_N = normal force and the force exerted on the surface is mainly due to the weight of the vehicle

F_f = contributed by the adhesion force at the tire–road interface as a result of molecular interactions, and the deformation force is mainly contributed by the viscous component of rubber in the tire.

The weight of the tire also has a role in F_f. The longitudinal force of the tire is generated when the vehicle is travelling in a straight line during acceleration or braking. The weight of the vehicle determines the vertical force (F_v) and depends on vehicle-related parameters.

The frictional force varies with the temperature under running conditions and the velocity change the tire is subjected to. In terms of the deformation frequency of the tire, there is an equivalence between deformation frequency and temperature.

Static friction acts when the tire is in contact with the road surface and remains stationary. As the applied force increases and surpasses the static friction threshold, it is overcome and kinetic friction takes over.

Other than the viscoelastic nature of a tire, one factor that comes into play while applying a brake is slip, described as the speed difference between the vehicle and the tire [36]. If a vehicle is traveling at 60 km/h and its tire rotates with a speed difference of 5 km/h, the tire is said to have a slip rate of 8.3%. Slip is a factor that allows the generation of braking force.

The grip of the tire while cornering can be related to the slip angle. Due to the slip angle, the tire slides slightly on the road surface, creating some friction needed for grip. If there is no slip angle, there is no lateral force and so no grip while cornering. The centrifugal force which develops is opposed by lateral forces as tires deform under lateral forces. This generates grip.

The importance of hysteresis of rubber is that the various aspects during the rolling of the tire, such as rolling resistance, the energy lost during deformation, the grip of the tire and resistance of the tire to skidding on different types of road surfaces, are all due to hysteresis.

In short, rubber exhibits grip because of its viscoelastic nature. When compared to synthetic rubbers like BR or SBR, NR offers superior road grip. This is because NR has a better ability for strain-induced crystallization at very high deformation frequencies, which leads to higher hysteresis and improved strength-related properties. The tire–road interaction leads to the generation of frictional forces, with the result that the tire remains in contact with the ground rather than skidding. For this, the tire needs to have a somewhat medium modulus and high hysteresis, so that the deformation is not instantaneous.

The deformation frequency applicable to rolling resistance is based on the speed of rotation of the tire. The information provided on a passenger car tire of size P215/65 R 15 is that the tire has a section width of 215 mm, an aspect ratio of 65, a rim diameter of 15 inches and 482 rev/km. At a driving speed of 100 km/h, the tire makes contact with the road roughly 13 times per second. So, the deformation frequency, to an extent depending on speed and road conditions, is in the range of 10^1 to 10^2 Hz. When allowance is made for road roughness at the micro and macro levels, the frequency range relative to road grip shifts into a higher deformation frequency domain, ranging from 10^4 to 10^6 Hz.

The rubber conforms to the micro-roughness of the surface and gradually releases from it, which is essential for maintaining good road grip. While the brake is applied the deformation frequency can range from 10^4 to 10^5 Hz. The frequency associated with wet skidding is high, about 10^3 to 10^6 Hz [35, 37].

Loss tangent, obtained from dynamic mechanical analysis, can be used to predict important properties, such as rolling resistance and grip. In the rolling conditions of a tire, the rolling resistance and road grip can be estimated from the values of loss tangent obtained at a frequency of 10^1 to 10^2 Hz and about 10^4 to 10^5 Hz. A low tan delta at a temperature of 60 °C under a frequency of 10 Hz correlates with low rolling resistance while a high tan delta at 0 °C correlates with good road grip. A temperature change (lowering of temperature) by 60 °C is equivalent to a frequency change (increase in frequency expressed in Hz) of about seven decades in the rubbery region of rubber deformation.

3.12 Effect of Method of Filler Addition on Dynamic Mechanical Properties

The dynamic properties of rubber are influenced by both the fillers and their method of incorporation. Adding fillers at the latex stage promotes better dispersion, which is further enhanced during mill mixing of other required ingredients, followed by vulcanization. When dual fillers like carbon black are silica are used there is also a possibility of interaction.

Data about fillers added to latex (masterbatch) and fillers added to the dry rubber on the mixing mill (control) are provided in Table 3.1, Table 3.2, Table 3.3 and Figure 3.11.

However, if the surfactant, which is fatty acid soap, is added to latex in higher concentration, it can affect the curing of rubber and this also will be reflected in the dynamic mechanical properties. It has previously been reported that palm oil fatty acid, like other well-known retarders, i.e., benzoic acid, retards the onset of vulcanization [38, 39]. An earlier study shows that the specific vulcanization rate is decreased if the concentration of stearic acid is increased [40, 41].

As a general observation, the dynamic mechanical properties related to elasticity are superior when fillers are added as dispersions into the latex. It is also observed from Figure 3.11 that the temperature at which the tan delta value is a maximum is marginally lower for the vulcanizates prepared from the fresh natural rubber latex-based masterbatch than for dry natural rubber-based vulcanizates. The reason may be the higher chain flexibility arising from to lower filler–filler interactions. Ideal compounds for tire tread possess high polymer–filler and low filler–filler interactions.

Table 3.1 Elastic Modulus and Tan Delta for Latex Masterbatch and Control Mixes Containing 25 phr HAF Carbon Black and 25 phr Silica at 10 Hz and 0.12% Strain

Properties	Sample 25/25 HAF/silica	
	Latex Masterbatch	Control (Dry Mill Mix)
Storage modulus (E'), Pa	3.05×10^7	1.53×10^7
Tan delta at 60 °C	0.08467	0.09146

Table 3.2 Loss Tangent at 10 Hz and 0.12% Strain for Latex Masterbatch and Control Mixes Containing 50 phr Reinforcing Fillers Consisting of HAF, ISAF and Silica at 60 °C

Sample Name	Latex Masterbatch		Control (Dry Mill Mix)
Fillers used	ISAF/silica 40/10	HAF/silica 25/25	HAF/silica 25/25
Tan delta at 60 °C	0.08627	0.08467	0.09146

Table 3.3 Loss Tangent Values at a Frequency of 11 Hz, Dynamic Strain of 0.025, Static Strain of 0.05, for Latex Masterbatch and Control Mixes Containing 45phr ISAF black and 50 phr ISAF/Silca fillers at 60 and 70 °C

s/n	Sample name	Loss tangent values			
		Masterbatch		Control (Dry Mill Mix)	
		60 °C	70 °C	60 °C	70 °C
1	45 phr ISAF black	0.2544	0.2467	0.2416	0.2323
2	40/5 phr ISAF black/silica	0.2451	0.2355	0.2661	0.2534

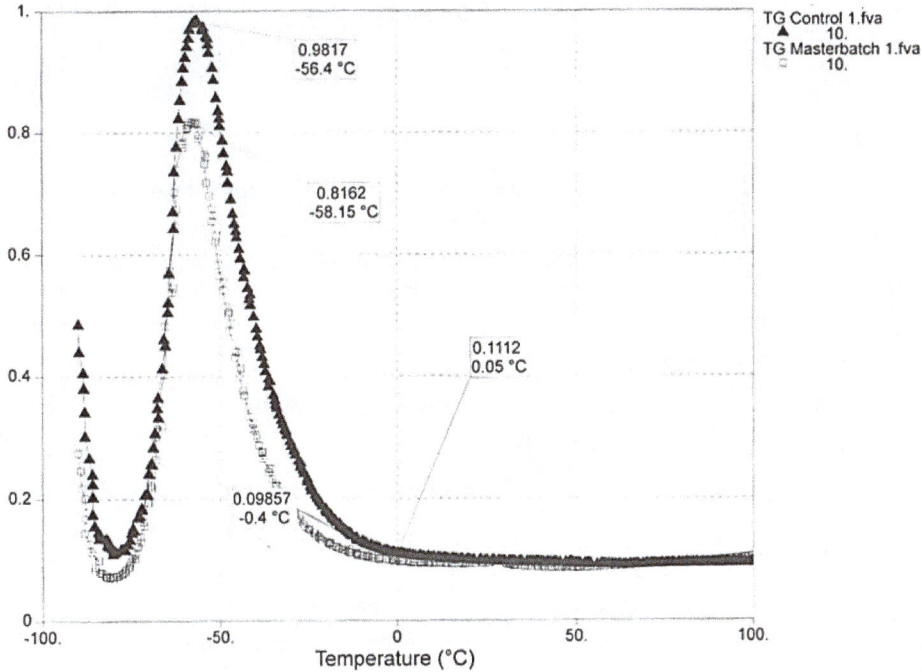

Figure 3.11 The road grip and rolling resistance of natural rubber vulcanizate containing 25 phr HAF carbon black and 25 phr silica incorporated at the dry stage and into natural rubber latex as dispersions

3.13 Service Life Prediction

The properties of a tire are predicted on the basis of the time-temperature superposition (TTS) principle. It is also important to know the service life of a product. During degradation of rubber, structural changes may result from physical or chemical transformations under the influence of long-term external factors, such as heat, ozone, oxygen, UV radiation, light radiation, chemical substances, water vapor, high-energy radiation and dynamic stresses which cause deterioration of the primary use properties. The main type of degradation of natural rubber is reaction with oxygen. The Arrhenius equation models this condition. Two assumptions underpin this application of Arrhenius's theory: (1) The mechanism of oxidation does not change in the different temperatures of accelerated aging, and (2) The oxidation proceeds uniformly throughout the material.

The temperature dependence of the rate constant is given by:

$$k = Ae^{-Ea/RT} \tag{3.82}$$

$$\ln k = -(Ea/R)(1/T) + \ln A \tag{3.83}$$

k = rate constant. It quantifies the rate and direction of a chemical reaction. If the concentration is in mol L^{-1} and the time is in seconds, then the units will be mol $L^{-1}s^{-1}$

A = frequency factor for collisions. It allows for the collisions between molecules and can be thought of as the number of correctly oriented collisions between reactant molecules that can lead to products

Ea = activation energy (J mol^{-1})

t = minimum energy required for reactants to transform into products

R = gas constant (8.3145 J $K^{-1}mol^{-1}$)

T = temperature in Kelvin

e = 2.718

Since aging results in the deterioration of general mechanical properties, such as tensile strength and elongation at break and tear strength, measuring the change in these properties during accelerated aging conditions is possible. The time required for retention of a certain percentage of tensile strength, say 55% of the original value at a particular temperature, can be taken as the rate of reaction at that temperature. Plots of ln k (rate of reaction) as a function of 1/T yields a straight line and, on extrapolation to room temperature, the rate of a reaction at room temperature (time taken for retention of 55% of the original value) is obtained.

Accelerated aging for mechanical properties is generally used for a large number of rubber products. One method used for gaskets and seals is that of compressive stress relaxation at constant temperature and is based on ISO 3384 (Rubber, Vulcanized or Thermoplastic – Determination of Stress Relaxation in Compression – Part 1: Testing at Constant Temperature).

The stress decay of polymer components under constant compressive strain is called compressive stress relaxation. This test measures the sealing force exerted by a seal or O-ring under compression between two plates. It provides definitive information for predicting the service life of materials by measuring the sealing force decay of a sample as a function of time, temperature and environment. It consists of stainless-steel compression devices. Force decay can be plotted against time to yield compressive stress relaxation. Under accelerated aging conditions, a specific threshold – such as a 25% reduction in stress within a defined time – can serve as a performance criterion.

Notations used

E	Young's modulus, elastic modulus
η	Viscosity
ER	Relaxation modulus
K	Boltzmann constant
A	Cross-sectional area
W	Work done
F	Stress per unit of cross-sectional area
S	Number of network chains per unit
Vf	Free volume
Vo	Occupied volume
Φ	Volume fraction of filler
V	Total volume
Fr	Fractional free volume
Frg	Fractional free volume at T_g
σ	Linear stress
ϵ	Linear strain
$\dot{\epsilon}$	Linear strain rate

References

[1]   K. P. Jones, P. W. Allen, 'Historical Development of the World Rubber Industry', in *Developments in Crop Science*, vol. 23, Elsevier, 1992, pp. 1–25. doi: 10.1016/B978-0-444-88329-2.50007-6.

[2]   P. J. George, C. Kuruvilla Jacob, Eds., *Natural rubber: agromanagement and crop processing*. Kottayam, India: Rubber Research Institute of India, 2000.

[3]   P. J. George, K. K. Thomas, A. O. N. Panicer, C. Kuruvilla Jacob, Eds., 'Indian Rubber Plantation Industry, Genesis and development', in *Natural rubber: agromanagement and crop processing*, Kottayam, India: Rubber Research Institute of India, 2000, p. 3.

[4]   B. Erman, J. E. Mark, C. M. Roland, *The science and technology of rubber*, 4th ed. Amsterdam Boston: Elsevier Academic Press, 2013.

[5]   C. M. Blow, S. H. Morell, 'The chemistry and technology of vulcanization', in *Rubber technology and manufacture*, London: Butterworths for the Institution of the Rubber Industry, 1971.

[6]   P. J. Flory, J. Rehner, 'Statistical Mechanics of Cross-Linked Polymer Networks I. Rubberlike Elasticity', *J. Chem. Phys.*, vol. 11, no. 11, pp. 512–520, Nov. 1943, doi: 10.1063/1.1723791.

[7]   J.-B. Donnet, *Carbon Black: Science and Technology*, 2nd ed. Boca Raton: CRC Press LLC, 1993.

[8]   A. Einstein, 'Eine neue Bestimmung der Moleküldimensionen', *Ann. Phys.*, vol. 324, no. 2, pp. 289–306, Jan. 1906, doi: 10.1002/andp.19063240204.

[9]   A. Einstein, 'Berichtigung zu meiner Arbeit: „Eine neue Bestimmung der Moleküldimensionen", *Ann. Phys.*, vol. 339, no. 3, pp. 591–592, Jan. 1911, doi: 10.1002/andp.19113390313.

[10] H. M. Smallwood, 'Limiting Law of the Reinforcement of Rubber', *J. Appl. Phys.*, vol. 15, no. 11, pp. 758–766, Nov. 1944, doi: 10.1063/1.1707385.

[11] A. I. Medalia, 'Effective Degree of Immobilization of Rubber Occluded within Carbon Black Aggregates', *Rubber Chem. Technol.*, vol. 45, no. 5, pp. 1171–1194, Sep. 1972, doi: 10.5254/1.3544731.

[12] C. M. Roland, 'Reinforcement of Elastomers', in *Reference Module in Materials Science and Materials Engineering*, Elsevier, 2016, p. B9780128035818021639. doi: 10.1016/B978-0-12-803581-8.02163-9.

[13] S. Wolff, 'Chemical Aspects of Rubber Reinforcement by Fillers', *Rubber Chem. Technol.*, vol. 69, no. 3, pp. 325–346, Jul. 1996, doi: 10.5254/1.3538376.

[14] H. A. Yasir, M. H. A. Maamori, H. M. Ali, 'Effect of Carbon Black Types on Curing Behavior of Natural Rubber', 2015.

[15] N. Tricás, E. Vidal-Escales, S. Borrós, 'The role of carbon black surface activity and specific surface area in the vulcanization reaction'.

[16] S.-J. Park, M.-K. Seo, C. Nah, 'Influence of surface characteristics of carbon blacks on cure and mechanical behaviors of rubber matrix compoundings', *J. Colloid Interface Sci.*, vol. 291, no. 1, pp. 229–235, Nov. 2005, doi: 10.1016/j.jcis.2005.04.103.

[17] C. G. Robertson, N. J. Hardman, 'Nature of Carbon Black Reinforcement of Rubber: Perspective on the Original Polymer Nanocomposite', *Polymers*, vol. 13, no. 4, p. 538, Feb. 2021, doi: 10.3390/polym13040538.

[18] W. Meon, A. Blume, H.-D. Luginsland, S. Uhrlandt, 'Silica and Silanes', in *Rubber Compounding*, B. Rodgers, Ed., CRC Press, 2004. doi: 10.1201/9781420030464.ch7.

[19] J. W. ten Brinke, S. C. Debnath, L. A. E. M. Reuvekamp, J. W. M. Noordermeer, 'Mechanistic aspects of the role of coupling agents in silica–rubber composites', *Compos. Sci. Technol.*, vol. 63, no. 8, pp. 1165–1174, Jun. 2003, doi: 10.1016/S0266-3538(03)00077-0.

[20] L. R. G. Treloar, The Physics of Rubber Elasticity, 3rd ed., Oxford University Press, 2005.

[21] E. M. Dannenberg, 'Molecular slippage mechanism of reinforcement', *Trans. Inst. Rubber Ind.*, vol. 42, p. 26.

[22] M. D. Stern, A. V. Tobolsky, 'Stress-Time-Temperature Relations in Polysulfide Rubbers', *J. Chem. Phys.*, vol. 14, no. 2, pp. 93–100, Feb. 1946, doi: 10.1063/1.1724110.

[23] A. V. Tobolsky, R. D. Andrews, 'Systems Manifesting Superposed Elastic and Viscous Behavior', *J. Chem. Phys.*, vol. 13, no. 1, pp. 3–27, Jan. 1945, doi: 10.1063/1.1723966.

[24] R. D. Andrews, A. V. Tobolsky, E. E. Hanson, 'The Theory of Permanent Set at Elevated Temperatures in Natural and Synthetic Rubber Vulcanizates', *J. Appl. Phys.*, vol. 17, no. 5, pp. 352–361, May 1946, doi: 10.1063/1.1707724.

[25] A. Wahab, A. S. Farid, 'Correlation between Compression-Set and Compression Stress-Relaxation of Epichlorohydrin Elastomers', *Polym. Polym. Compos.*, vol. 19, no. 8, pp. 631–638, Oct. 2011, doi: 10.1177/096739111101900802.

[26] M. D. Ellul, E. Southern, 'Comparison of Stress Relaxation with Compression Set for Seal Compounds', *Plast. Rubber Compos. Process. Appl.*, vol. 5, no. 1, pp. 61–69, 1985.

[27] 'IS 3400-11 (1985): Methods of test for vulcanized rubbers, Part 11: Determination of rebound resilience'.

[28] L. Mullins, 'Effect of Stretching on the Properties of Rubber', *Rubber Chem. Technol.*, vol. 21, no. 2, pp. 281–300, Jun. 1948, doi: 10.5254/1.3546914.

[29] J. A. C. Harwood, A. R. Payne, 'Stress softening in natural rubber vulcanizates. Part III. Carbon black-filled vulcanizates', *J. Appl. Polym. Sci.*, vol. 10, no. 2, pp. 315–324, Feb. 1966, doi: 10.1002/app.1966.070100212.

[30] L. Mullins, N. R. Tobin, 'Stress softening in rubber vulcanizates. Part I. Use of a strain amplification factor to describe the elastic behavior of filler-reinforced vulcanized rubber', *J. Appl. Polym. Sci.*, vol. 9, no. 9, pp. 2993–3009, Sep. 1965, doi: 10.1002/app.1965.070090906.

[31] W. Fu *et al.*, 'Mechanical Properties and Mullins Effect in Natural Rubber Reinforced by Grafted Carbon Black', *Adv. Polym. Technol.*, vol. 2019, pp. 1–11, Aug. 2019, doi: 10.1155/2019/4523696.

[32] A. R. Payne, 'The dynamic properties of carbon black – loaded natural rubber vulcanizates. Part I', *J. Appl. Polym. Sci.*, vol. 6, no. 19, pp. 57–63, Jan. 1962, doi: 10.1002/app.1962.070061906.

[33] J. D. Ferry, *Viscoelastic properties of polymers*, 3rd ed. New York Chichester Brisbane Toronto Singapore: John Wiley & Sons, 1980.

[34] M. L. Williams, R. F. Landel, J. D. Ferry, 'The Temperature Dependence of Relaxation Mechanisms in Amorphous Polymers and Other Glass-forming Liquids', *J. Am. Chem. Soc.*, vol. 77, no. 14, pp. 3701–3707, Jul. 1955, doi: 10.1021/ja01619a008.

[35] C. M. Roland, 'GLASS TRANSITION IN RUBBERY MATERIALS', *Rubber Chem. Technol.*, vol. 85, no. 3, pp. 313–326, Sep. 2012, doi: 10.5254/rct.12.87987.

[36] K. C. Ludema, 'Physical factors in tire traction', *Phys. Technol.*, vol. 6, no. 1, pp. 11–17, Jan. 1975, doi: 10.1088/0305-4624/6/1/302.

[37] R. R. Rahalkar, 'Dependence of Wet Skid Resistance upon the Entanglement Density and Chain Mobility according to the Rouse Theory of Viscoelasticity', *Rubber Chem. Technol.*, vol. 62, no. 2, pp. 246–271, May 1989, doi: 10.5254/1.3536243.

[38] B. A. Dogadkin, A. V. Dobromyslova, O. N. Belyatskaya, 'The Scorching of Rubber Mixes. II. Effect of Retarders on the Kinetics of Sulfur Combination', *Rubber Chem. Technol.*, vol. 35, no. 2, pp. 501–508, May 1962, doi: 10.5254/1.3539922.

[39] H. Ismail, T. A. Ruhaizat, 'Effect of Palm Oil Fatty Acid on Curing Characteristics and Mechanical Properties of CaCO3 Filled Natural Rubber Compounds', *Iran. Polym. Journal*, vol. 6, no. 2, 1997.

[40] A. Y. Coran, 'Vulcanization. Part VII. Kinetics of Sulfur Vulcanization of Natural Rubber in Presence of Delayed-Action Accelerators', *Rubber Chem. Technol.*, vol. 38, no. 1, pp. 1–14, Mar. 1965, doi: 10.5254/1.3535628.

[41] A. Y. Coran, 'Vulcanization. Part V. The Formation of Cross-links in the System: Natural Rubber-Sulfur-MBT-Zinc Ion', *Rubber Chem. Technol.*, vol. 37, no. 3, pp. 679–688, Jul. 1964, doi: 10.5254/1.3540360.

[42] L. R. G. Treolar, 'The physics of Rubber Elasticity' 3rd Edition, p. 3, Clarendon Press Oxford, 2005

4

Basic Principles of Vibration, Shock and Sound Isolation Based on Rubber Viscoelasticity

Rosamma Alex, Pradeepkumar P. Joy, Baby Kuriakose

4.1 Introduction

In our daily lives, we frequently witness energy transfer between systems – for example, when operating mechanical equipment, driving vehicles or cooking food. When one body moving at a particular velocity comes in contact with another, it transfers its kinetic energy to that body. If the system has the ability to store energy, it will start vibrating at a particular frequency. If the force is an impulse force, the body experiences a sudden change in acceleration in a period of milliseconds and the body will not vibrate. Rather, a sudden, instantaneous motion happens, called shock. In conditions of shock, there is an instantaneous transfer of kinetic energy. Shock is a transient high-energy event that occurs when an object experiences very high acceleration or deceleration, such as during a hammer strike or a high-speed vehicle collision. This sudden motion causes displacements that disturb the surrounding air molecules, causing them to collide and generating pressure fluctuations. These pressure waves propagate through the air, converting mechanical energy into sound energy. Vibration shock and sound need to be attenuated and this can be achieved to a satisfactory limit through the use of rubber.

Vibrations have been of interest for a long time. One fascinating area of vibration is related to music and musical instruments that produce pleasant sounds, due to vibration of the instrument's membranes, surfaces or strings. But not all vibrations are pleasant. The vibrations made by machines are unpleasant and harmful to the equipment. Moreover, these vibrations also make the foundation vibrate, leading to their failure. A notable feature of equipment such as electric motors, automobile engines and generators is that they cause unwanted sound and vibrations when in service.

Vibration is an oscillatory motion in mechanical components or systems with elasticity when they are acted on by some force and disturbed from their equilibrium position. The component becomes displaced by the force applied and the opposing force, in the form of elastic potential energy, tries to bring the body back to its original position.

Vibrations arise in the ground as a result of natural disasters, such as earthquakes, landslides and floods, and are harmful to structures such as bridges, buildings and sophisticated equipment, often rendering them unusable. Delicate machines, including surgical microscopes, electronic equipment, lasers, MRI units, scanning electron microscopes and computer disk drives, can be damaged by external vibrations. Living beings, too, are damaged.

Vibrations, unwanted sounds and shock must be reduced to protect human life and other mechanical structures. Rubber pads suitably bonded to metals are simple devices that can easily reduce the transmission of vibrations arising from various equipment to the support and ground or from ground to structures.

Natural rubber is used in engineering applications because of its unique viscoelastic nature. Vulcanized rubber has the ability to store energy by virtue of its elasticity and also its ability to dissipate energy by virtue of hysteresis. Suitably compounded NR vulcanizates exhibit high elasticity, along with tensile strength of about 30 MPa and elongation of up to about ten times the original length. As observed in dynamic mechanical analyses, rubber exhibits a high storage modulus with low values of loss tangent. The unique micro- and microstructure, combined with a suitable level of reinforcing fillers, can offer the required energy absorption or material damping characteristics, as measured by loss tangent. The elasticity and hysteresis are judiciously balanced in applications such as shock and vibration isolators. The viscoelasticity of rubber also contributes to the absorption of sound. Being a porous medium with interconnected cells, natural rubber latex-based foam or dry rubber-based sponge acts as an effective sound absorber by dissipating sound energy through collisions of the air molecules within the pores.

The flexibility and viscoelastic properties of rubber and its ability to be molded into porous structures make it an effective material for sound absorption.

Rubber mounts reduce the transmission of vibration from a source to other machines, machine parts or the support structure. Vibration isolation by rubber mounts is strongly related to the frequency ratio of input vibration and natural frequency of mount and resilient rubber mounts effectively isolate a structure from the transmission of vibration.

Rubber mounts reduce shock transmission by absorbing kinetic energy. The force of a shock can also cause the mount to start vibrating at its natural frequency and this affords a way to isolate shock with a rubber mount.

Mathematical expressions for the transmission of vibration to a structure are based on forces encountered in the process. Shock severity can be characterized by plotting acceleration against time. Such plots can have different geometrical shapes. The intensity of the shock, as measured by velocity changes, can be quantified by integrating the area enclosed by the shock pulse, that is, under the acceleration–time curve.

Damping in rubber mounts arises from energy loss due to hysteresis. The mathematical relation for energy loss is based on the two components of viscoelasticity in rubber: the elastic component and the viscous component, the latter being out of phase during sinusoidal deformation.

The effectiveness of vibration reduction, as measured by the transmissibility of force or motion, to a system is derived from the solution of the general differential equation governing forced damped vibrations. The displacement of mass in such systems, influenced by elastic and viscous forces, is governed by the damping efficiency, quantified by a parameter called the damping ratio. In the design of rubber mounts, energy loss is attributed to material damping, which arises from internal friction due to molecular motion during deformation. This form of damping is conceptually related to viscous damping, which involves energy dissipation through fluid flow resistance.

4.2 Vibratory Systems

In general, a system with mass and elasticity can be called a vibratory system and has four mechanical components: the kinetic-energy-storing device (mass), the potential-energy-storing device (a spring associated with stiffness), the energy-dissipating-device (damper) and a source for excitation of the system. Energy dissipation also called damping can happen in vibratory systems by viscous damping, dry friction damping (Coulomb damping) which occur through friction between moving parts or by hysteretic damping (material damping) which occur through internal friction within the material. Two significant energy dissipation process used to control vibration is by viscous and hysteretic damping. In many vibratory systems, there can also be some unbalanced force. A schematic diagram of a vibratory system is shown in Figure 4.1.

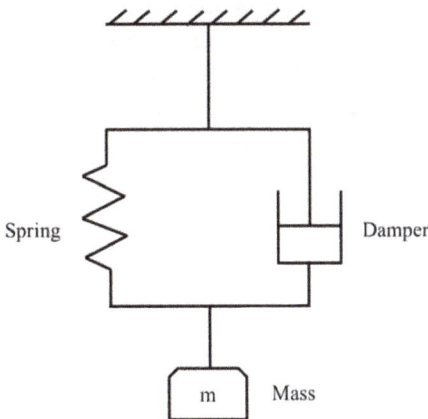

Figure 4.1 Components of a vibratory system

With these mechanical components, a vibratory system can execute three types of vibration: free or natural, forced, and damped. In a vibratory system with no unbalanced force or friction, the vibration executed by the body without any external force acting on it after the initial displacement is called the free or natural vibration. The frequency of such a vibration is called the natural frequency. The amplitude of such a vibration decreases with time, due to the effect of external damping factors. When a body vibrates under the influence of external force, the vibration is called forced vibration and the vibration frequency will be that of the external vibration. The amplitude of vibration will also remain constant over time. When there is a reduction in vibration amplitude in every vibration cycle as a result of damping forces present in the system, the vibration is called damped vibration.

4.3 Forced Vibration

Vibrations in a system that are induced by external force or external excitation and may be periodic or random are called forced vibrations. If the external vibration is periodic, the system will vibrate with periodic vibrations of the same frequency as the external excitation. Suppose the frequency of external vibration becomes the same as the natural frequency of the system; resonance occurs and the amplitude of oscillation of the system becomes very high. Examples of forced vibrations are the vibration of a vehicle due to up and down movements arising from road undulations, the shaking of mixing equipment as a result of force imbalance and the vibration of a building, visualized as side-to-side movement, in an earthquake. Many of the vibrations experienced in our lives are forced vibrations.

The external excitation leading to the forced vibration of a system may be a constant harmonic force, transient and non-harmonic force or non-periodic and non-harmonic force. For example, the vibration experienced by the floor on which a motor rests may be a constant harmonic force stemming from the motor's operation. In the case of shock pulses, external force is regarded as transient and non-harmonic, accompanied by very high acceleration and short duration. Earthquake forces are non-periodic and non-harmonic, characterized by very high acceleration of short duration, arising from the ground and causing lateral vibrations in buildings and other structures.

4.4 Free Vibration

The simplest form of periodic force is sinusoidal and mathematical relations can be derived from this type of vibration.

4.4.1 Mathematical Expression for the Natural Frequency of a Free-Vibration System

In a free-vibration system, there is no damping or unbalanced force. Consider a spring-mass system attached to an immovable support, a simple vibration system on a frictionless surface with a single degree of freedom (Figure 4.2). In the absence of force, the spring is at rest or equilibrium.

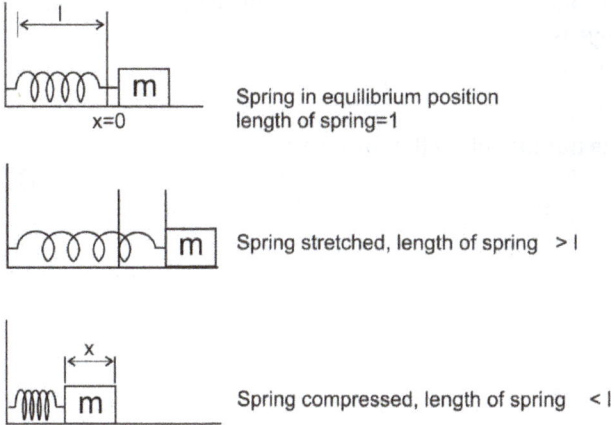

Spring in equilibrium position
length of spring=1

Spring stretched, length of spring > l

Spring compressed, length of spring < l

Figure 4.2 Spring-mass system in three conditions: spring in equilibrium, spring stretched, and spring compressed, with periodic vibration of spring-mass system and displacement X

If a force (F) acts on the spring-mass system in the horizontal direction, the spring either compresses or elongates. Since the spring is elastic, it opposes the applied external force with a force developed in the spring called the restoring force. This restoring force is proportional to the displacement (X) (or deflection) and a linear relationship exists between the restoring force and the spring's deflection.

This behavior is described by Hooke's law:

$$F \alpha X$$

$$F = kX \tag{4.1}$$

Where

k = spring constant or dynamic spring stiffness

In free vibration, the mass is given an initial displacement and no external force acts on the system during its motion. The spring stores the energy, which is used by the mass to oscillate about the mean position. Periodic motion of this kind can be represented by a sine function. The position of the mass as a function of time ($X(t)$) is given by:

$$X(t) = A\ sin(\omega t) \tag{4.2}$$

Where

ω = angular frequency or angular displacement of mass per unit time

A = maximum amplitude of oscillation

The rate at which one cycle of rotation is completed is called the angular velocity. For one complete rotation, the angle is 2π radians and the taken is T, so:

$$\omega = 2\pi/T$$

The frequency f, defined as the number of oscillations per unit time, is:

$$f = 1/T$$

From this, we also get:

$$T = 2\pi/\omega$$

Or

$$\omega = 2\pi f$$

The velocity of the object (mass) as a function of time is the derivative of displacement:

$$v(t) = dX(t)/dt = -\omega A\ cos(\omega t) \tag{4.3}$$

The acceleration is the derivative of velocity:

$$a(t) = dv(t)/dt = -\omega^2 A\ sin(\omega t) = -\omega^2 X \tag{4.4}$$

Thus, in simple harmonic motion, acceleration is proportional to displacement and opposite in direction:

$$a = -\omega^2 X$$

Applying Newton's Law ($\Sigma\ F = 0$), the differential equation for natural vibration becomes:

$$m\ddot{X} + kX = 0 \tag{4.5}$$

$$-m\omega^2 X = -kX$$

This leads to:

$$\omega^2 = k/m$$

Replacing ω by ω_n we arrive at the natural angular frequency:

$$\omega_n = \sqrt{k/m} \tag{4.6}$$

Assuming that the spring has negligible mass, the natural frequency of a spring-mass system in terms of frequency is:

$$f_n = \frac{1}{2\pi}\sqrt{\frac{k}{m}} \tag{4.7}$$

When expressed in terms of static deflection, δS, where $(k / \delta S) = mg$, the natural frequency becomes:

$$f_n = \frac{1}{2\pi}\sqrt{\frac{g}{\delta S}} = 3.13\sqrt{\frac{1}{\delta S}} \tag{4.8}$$

The spring stiffness is given by:

$$k = F/\delta S = mg/\delta S \tag{4.9}$$

Here, δS is the static deflection (the deflection of the spring due to the mass), typically expressed in inches or millimeters. The gravitational acceleration g is $g = 980.665$ cm/s^2 = 386.093 in/s^2 = 32.1739 ft/s^2.

Using g as 386.093 in/s^2, the natural frequency simplifies to:

$$f_n = 3.13\sqrt{\frac{1}{\delta S}} \tag{4.10}$$

In undamped free vibration, as illustrated in Figure 4.3, the amplitude of oscillation remains constant across successive cycles of vibration. However, vibration typically stops after a certain number of cycles, as a result of internal or external damping forces.

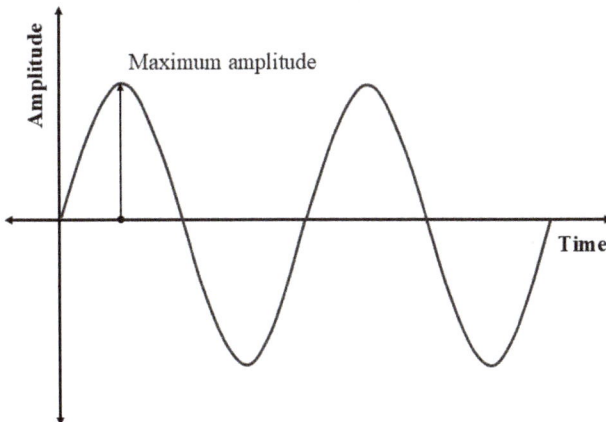

Figure 4.3 Undamped free vibration

4.4.2 Free Damped Vibration

In a vibratory system, there can be a loss of energy that causes the vibration to stop, and this is called damping. Damping can be described mainly in different ways: viscous damping due to fluid resistance and material damping due to hysteresis of the material.

4.4.2.1 Viscous Damping

There are different types of damping: coulomb, viscous and hysteretic. Coulomb damping arises from friction between two dry surfaces that rub against each other [1]. Viscous damping occurs when an object moves through a fluid, with the damping force being proportional to the object's velocity. A common example is an automotive shock absorber, where fluid is forced through narrow openings in the piston. This flow resistance generates damping forces and frictional losses occur within the fluid as a result of the piston's motion.

4.4.2.1.1 Mathematical Expressions to Quantify Viscous Damping

In vibration isolation, the two types of damping are more significant, i.e., material and viscous damping.

Viscous damping is related to the velocity gradient generated during fluid flow. Mathematically, the damping force is related to the variation in velocity within the fluid as it undergoes deformation. Viscous damping is characterized by two parameters: the damping coefficient and the damping ratio. It can be observed in the vibration of a spring-mass system in a fluid, where the vibration amplitude gradually decreases slowly and eventually stops, due to the damping force (Fd) generated in the fluid. The damping force depends on the velocity gradient between the fluid layers.

The relationship between damping force and amount of damping can be represented mathematically using a dashpot (Figure 4.4). This is a device that consists of a cylinder filled with fluid and a vibrating element attached to a piston that moves within it. The dashpot resists motion and is used to stop vibrations.

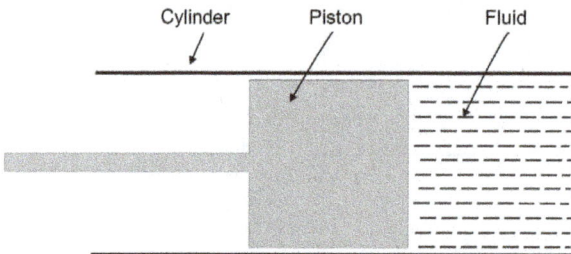

Figure 4.4 Schematic representation of a dashpot or damper

In a dashpot (Figure 4.4), when a force is applied to the piston, it moves over the surface of the thin fluid layer enclosed in the space between the piston and the cylinder wall. This force causes the liquid layer adjacent to the piston to move with velocity (v), while the layer in contact with the stationary cylinder wall remains at rest. The thickness of the fluid layer (y) and the curved surface area of the piston (A) over which the force is applied are known. The damping force in the dashpot is given by:

$$Fd = \mu A v / y \tag{4.11}$$

Where

μ = viscosity of the fluid

Assuming $\mu A/y$ is constant, the damping force becomes directly proportional to velocity:

$$Fd = \text{constant} \times v$$

If the displacement of the fluid layer is denoted by X, then the damping force can be expressed as:

$$Fd = c\, dX/dt \tag{4.12}$$

Or

$$Fd = c\dot{X}$$

Where

c = constant related to damping and is called the damping coefficient

4.4.2.1.2 Damping Ratio in Viscous Damping

Mathematical relations for various parameters that quantify viscous damping can be derived from a vibratory system consisting of a spring and damper, as shown in Figure 4.1.

When a force (F) acts vertically on the spring-mass–damper system (as shown in Figure 4.1) and is then removed, the system undergoes free-damped vibration. If a spring only were present, the body would exhibit simple harmonic motion, assuming no damping force. However, in this case, a damper is also present. The dashpot, modeled as a cylinder filled with fluid and fitted with a movable piston resists motion when a force is applied. As the piston moves through the fluid with velocity v, a damping force is generated that is proportional to this velocity.

In this spring–damper system, any applied force is opposed by the spring force and the damping force. The spring force is generated by the spring stiffness while the damping force arises due to the velocity of the piston, which is influenced by the fluid flow in the gap between the piston surface and cylinder wall.

According to Newton's second law, the net force acting on a vibratory system is zero when the applied force is balanced by the opposing forces: $\Sigma F = 0$.

In the presence of a damper, Equation 4.5 gains an additional term related to the velocity of the piston, which is opposed by damping forces generated by the fluid in the dashpot:

$$ma = -cv - kX \tag{4.13}$$

Or, expressed as a differential equation, the motion of damped free vibration, assuming the mass has a velocity and acceleration due to the applied force on the spring-mass system, is:

$$m\ddot{X} + c\dot{X} + kX = 0 \tag{4.14}$$

This is a second-order linear differential equation describing the motion of the mass in this vibratory system. The displacement of X as a function of time t can be determined by solving this equation.

A general solution for the displacement (X) of a mass in damped free vibration is obtained by assuming an exponential form. This transforms the differential equation into a corresponding quadratic equation:

$$X(t) = Xe^{\omega t}$$

Here, angular displacement is expressed as the product of angular velocity (ω) and time (t).

Substituting this assumed solution into Equation 4.14 yields the quadratic equation:

$$Xe^{\omega t}(m\omega^2 + c\omega + k) = 0 \tag{4.15}$$

This equation explains how the nature of damping in the system – whether it is critically damped, underdamped or overdamped. Overdamped denotes a vibration that comes very slowly to equilibrium. An example of an overdamped system is a heavy door equipped with a door closer (a spring–damper system) that closes gradually after being opened.

Critically damped denotes a condition in which vibration stops in the shortest-possible time without further vibration. An example of critical damping is a car shock absorber, which uses viscous damping to resist motion through hydraulic fluid flowing through a small orifice. The shock absorber damps the force exerted on the car due to its viscous-damping ability. A typical diagram of a viscous damper is shown in Figure 4.5.

Underdamped refers to a condition where the system continues to oscillate, but with decreasing amplitude until the vibration stops. For example, when a tuning fork is struck against a surface and removed, it begins to vibrate. Over time, these vibrations gradually decrease and eventually cease.

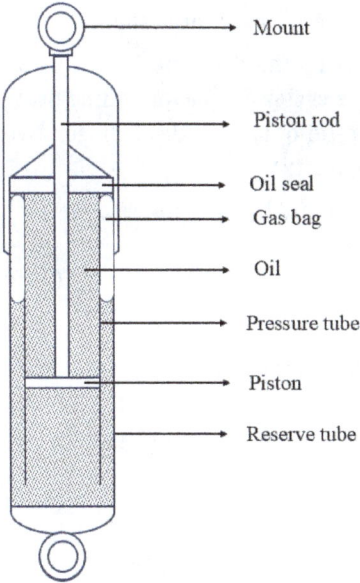

Mount

Piston rod

Oil seal
Gas bag

Oil

Pressure tube

Piston

Reserve tube

Figure 4.5 Diagrammatic representation of a twin-tube shock absorber

One solution to this equation involves a term called the critical damping coefficient (Cc). To an extent depending on the conditions, the damping coefficient is equal to, greater than or less than the critical damping coefficient (Cc).

(The two roots of the quadratic equation are $\frac{-c\pm\sqrt{c^2-4mk}}{2m}$)

The critical damping coefficient Cc is the value of c that makes the expression under the square root equal to zero; this means that, if $c = \sqrt{4mk}$ or $2\sqrt{km}$, the system is critically damped. Alternatively, the damping coefficient required for critical damping (Cc) is given by

$$Cc = 2\sqrt{km} \tag{4.16}$$

Using the natural angular frequency, the critical damping coefficient can be rewritten as:

$$Cc = 2\,m\omega_n \tag{4.17}$$

A system is critically damped and the vibratory system has enough damping to reach its resting position quickly without any vibration. When the damping coefficient, c, is small or $c < \sqrt{4mk}$, the system oscillates and the amplitude of oscillation decays exponentially with time. In this condition, the system is said to be underdamped. If the damping coefficient is very large or $c > \sqrt{4mk}$, the system is over-damped, in which case the vibratory system experiences a greater damping force than the force required for natural frequency, and so the system slowly comes to rest without vibrating (Figure 4.6).

A dimensionless number, called the damping ratio ζ is defined as:

$$\zeta = \text{Actual damping coefficient/critical damping coefficient} = \frac{c}{2m\omega_n} \tag{4.18}$$

The damping ratio characterizes the viscous damping capacity of the system.

The damping ratio can also be determined by analyzing the decrease in oscillation amplitude. The decrease in amplitude over successive cycles can be quantified using the logarithmic decrement δ, which represents the ratio of amplitudes between two successive peaks in a vibration cycle (refer to the Appendix).

The damping ratio ζ is related to the logarithmic decrement by the expression:

$$\zeta = \frac{\delta}{\sqrt{2\pi^2 + \delta^2}} \tag{4.19}$$

If the value of ζ is <1, the system is underdamped; if it is equal to 1, the system is critically damped; if it is >1, the system is overdamped.

The damping coefficient C is related to the critical damping coefficient Cc by:

$$C = Cc\,\zeta \tag{4.20}$$

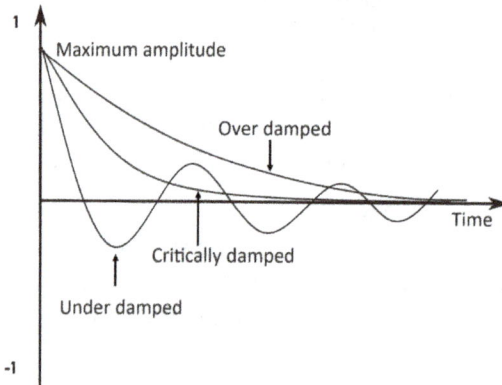

Figure 4.6 Damped vibrations, underdamped, critically damped and overdamped

4.4.3 Material Damping: Mathematical Equation for Quantifying Damping in Terms of Energy Loss

Material damping can be understood as the time lag between stress and strain, observed as a phase lag during the oscillatory deformation of rubber and typically measured with a dynamic mechanical analyzer. In the cyclic deformation of a rubber sample, stress (σ) lags behind strain (ϵ) by a phase angle (δ). It is assumed that the modulus of the sample during cyclic deformation consists of an elastic component and a viscous component. The viscous component, which is out of phase with the elastic component, is responsible for damping. In cyclic deformation, damping is associated with energy loss within the rubber, due to internal molecular friction.

During one deformation cycle, the area under the stress–strain curve during loading represents the work done by the elastic component. During unloading, some energy is lost; this energy loss can be described by a hysteresis loop. The hysteresis loop for rubber under cyclic deformation is schematically shown in Figure 4.7.

The work done W_1 in stressing the material (per unit volume), where ε is strain and σ is the stress, in one deformation cycle is given by:

$$W_1 = \int \sigma d\varepsilon \tag{4.21}$$

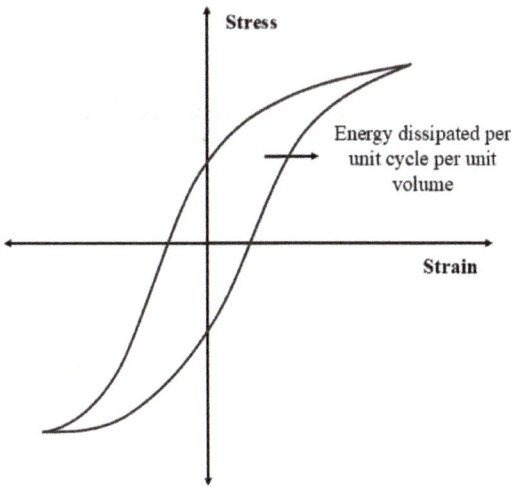

Figure 4.7 Schematic representation of energy loss in one cycle of deformation during cyclic deformation of the rubber sample

For a sinusoidal deformation of rubber, stress and strain can be described as follows:
Strain lags behind stress by a phase angle δ.
Strain:

$$\varepsilon t = \varepsilon_0 \sin \omega t$$

Strain rate:

$$\dot{\varepsilon} = \omega \varepsilon_0 \cos \omega t$$

Stress:

$$\sigma_t = \sigma_0 \sin(\omega t + \delta)$$

The work done per cycle $W1$ is given by:

$$W_1 = \int_0^T \sigma d\varepsilon \tag{4.22}$$

$$W_1 = \int_0^T \sigma \left(\frac{d\epsilon}{dt}\right) dt$$

Substituting the expressions for stress and strain rate:

$$W_1 = \omega \sigma_0 \epsilon_0 \int_0^T \sin(\omega t + \delta) \cos(\omega t) dt \tag{4.23}$$

(See Appendix)

Solving the integral yields:

$$W_1 = \pi \sigma_0 \, \epsilon_0 \, \sin \delta \tag{4.24}$$

The energy dissipated during a deformation cycle in terms of loss modulus is:

$$W_1 = \pi E'' \, \epsilon_0^2 \tag{4.25}$$

Where

E'' = loss modulus

Note:

The strain energy in the form of work done during extension is given by S.E. $= \frac{1}{2} E \epsilon^2$ (Equation 3.32), where E is the elastic modulus and ϵ is the strain, defined as the change in length of a strip (x) over the original length (l), i.e. ($\epsilon = x/L$). Part of the stored energy is lost during the unloading cycle.

The elastic energy stored in a cycle of deformation is:

$$W_2 = 1/4 \, E' \epsilon_0^2 \tag{4.26}$$

Where

E' = elastic modulus and ϵ_0 is the strain amplitude

The loss factor is defined as the ratio of energy dissipated to energy stored in a cycle (W_1/W_2). For rubbery materials, loss factor (mechanical loss) is the ratio of the viscous modulus to the elastic modulus.

$$\text{Loss tangent or } \tan \delta \text{ or mechanical loss} = \frac{E''}{E'} \tag{4.27}$$

Material damping is related to displacement amplitude, unlike viscous damping, which is related to displacement velocity. Therefore, identical values of damping ratio and loss tangent do not necessarily imply the same damping capacity. From practical observations, viscous damping is more effective than material damping under resonance conditions. (For instance, a dashpot with a particular damping coefficient typically provides more damping than a rubber pad with the same loss tangent under resonance conditions.) For practical purposes, loss tangent and damping ratio can be related in the form:

$$E''/E' = 2\zeta = \tan \delta \tag{4.28}$$

4.5 Vibration Isolation

Vibration isolators are simple devices that protect against external excitations. While they may not completely eliminate vibrations, they significantly reduce the transmission of vibration in terms of force and motion. This process helps safeguard machines against excessive incoming vibrations or prevents vibrations from being transferred from a machine to the foundation.

Rubber has both elastic and damping characteristics and so can act as a vibration isolator. It is particularly useful for isolating input vibrations of very high frequency. Other materials used for vibration isolation are cork, felt and mechanical springs. Metallic springs are useful for applications involving heavy loads and high frequencies.

Amongst these materials, rubber pads are widely favored because they are inexpensive, readily available, easy to process and highly efficient at isolating vibration.

Vibration isolators are installed between the machine and its foundation to reduce the amount of vibration transmitted to the foundation, whether in the form of motion or force. They function by absorbing, storing and dissipating part of the vibratory energy. Their effectiveness is primarily governed by two key properties: dynamic stiffness and damping.

4.5.1 Mathematical Expression for Transmissibility in Forced Damped Vibration Isolation

The vibratory system associated with vibration isolation of a foundation from rotor-induced vibrations can be modeled as forced damped vibration. Its components, consisting of a mass, spring with stiffness k, damper with damping coefficient c and external force $F_0 \sin \omega t$, are illustrated in Figure 4.8.

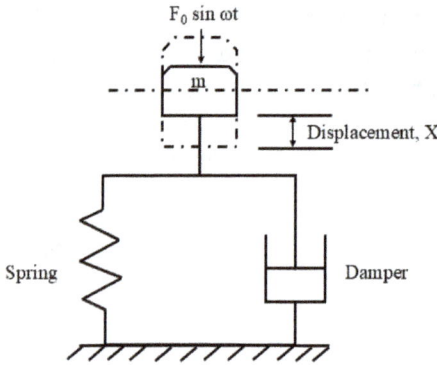

Figure 4.8 Forced damped vibratory system

In forced-damped vibratory systems, the forces present are the spring force, inertia force, friction force and external force generated by the operation of the machine's rotor. These forces can be represented using a general differential equation:

$$F_0 \sin \omega t = m\ddot{X} + c\dot{X} + kX \tag{4.29}$$

A schematic representation of vibration isolation using a vibration isolator is given in Figure 4.9.

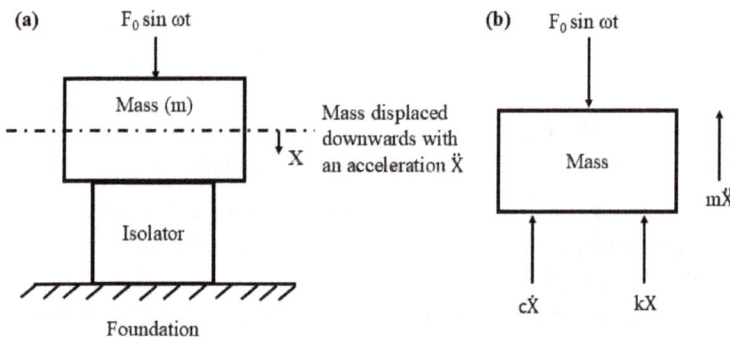

Figure 4.9 Mass–vibration isolator and system and free body diagram showing the different forces

The rubber isolator can be modeled as a combination of a spring with stiffness k and a damper with damping coefficient c. When the machine (represented by mass m) operates, it generates vibrations that result in displacement of the mass. When the machine is operating, a constant harmonic force ($F_0 \sin \omega t$) generated inside the machine causes vibration and displacement of the mass (X). This force can be transmitted to the foundation.

The level of protection given to the foundation is quantified by the transmissibility, expressed as the ratio of input to output, in terms of either force or motion. If the vibration source is a machine, transmissibility is the ratio of output force to input force.

Conversely, if the vibration source is the oscillatory motion of the foundation (motion excitation), transmissibility is defined as the ratio of the equipment's vibration amplitude to the foundation's vibration amplitude.

Transmissibility (Tr) in terms of force,

Tr = Force transmitted to the foundation (Ft)/Force impressed on the system (F_0)   (4.30)

This vibratory system has an input force in the form of harmonic motion. To oppose this excitation, three forces come into play: elastic, viscous and inertial. They act in different directions. In harmonic motion, displacement, velocity and acceleration are out of phase by 90°, as shown below.

The displacement (X) of the mass at any time t is:

$$X(t) = A \sin \omega t \tag{4.31}$$

Where

ω = circular frequency

A = amplitude

The velocity is given by:

$$v = \dot{X} = A\omega \, \cos \omega t = \omega A \sin(\omega t + 90^0) \tag{4.32}$$

The acceleration is given by:

$$a = \ddot{X} = -A\omega^2 \sin \omega t = \omega^2 A \sin(\omega t + 180^0) \tag{4.33}$$

Using a vector approach based on the solution for the differential equation for the motion of damped free vibration, the maximum displacement, velocity and acceleration can be expressed as:

$$\vec{X} = A \, e^{i\omega t} \tag{4.34}$$

Where

ω = circular frequency

A = amplitude

$$\frac{d\vec{X}}{dt} = \frac{d}{dt}(Ae^{i\omega t}) = i\omega Ae^{i\omega t} = i\omega \vec{X} \tag{4.35}$$

$$\frac{d^2\vec{X}}{dt^2} = \frac{di\omega \, Ae^{i\omega t}}{dt} = \omega^2 \vec{X}$$

The three forces acting on the vibratory system are:

$$F_1 = m\omega^2 X$$

$$F_2 = kX$$

$$F_3 = c\omega X$$

Here, k is the spring constant of the spring and X is the maximum displacement of the spring.

The damping force (related to the velocity) acts perpendicularly to the spring force while the viscous force acts perpendicularly to the inertia force (related to the acceleration). X is the maximum displacement in the machine and is the maximum amplitude during vibration. The displacement vector lags behind the impressed force an angle by φ. The damping force has a phase difference $\pi/2$ with the spring force. The angle subtended by spring force and the reference axis is ωt-φ.

From Figure 4.10 (a), it can be seen that the impressed force F_0 causes a mass displacement, but this lags behind the impressed force by an angle φ. Spring force (F_2) arises from the elastic component and the damping force (F_3) arises from the viscous component opposing the displacement. The spring force and damping force are 90° out of phase. The inertia force (F_1) is in phase with the displacement. The relative positions and magnitudes of these vectors do not change over time.

In the vector diagram (Figure 4.10 (b)), the input force can be represented as the hypotenuse of a right-angled triangle (ABC). The resultant input force is determined using Pythagorus' theorem.

$$F_0 = \sqrt{(kX - m\omega^2 X)^2 + (c\omega X)^2} \tag{4.36}$$

$$F_0 = X\sqrt{(k - m\,\omega^2)^2 + (c\omega)^2}$$

$$X = F_0/\sqrt{(k - m\,\omega^2)^2 + (c\omega)^2} \tag{4.37}$$

$$\frac{X}{F_0} = 1/\sqrt{(k - m\omega^2)^2 + (c\omega)^2} \tag{4.38}$$

on dividing throughout by k and by suitable substitution for ζ, and $\omega/\omega\,n$ we get

$$\frac{X}{F_0} = 1/k/\sqrt{(k/k - m\,\omega^2/k)^2 + (c\omega/k)^2}$$

Amplitude of vibration of the mass in the system $X = \dfrac{F_0/k}{\sqrt{(1-(\omega/\omega n)2)^2+(2\zeta\omega/\omega n)^2}}$

(a)

(b)

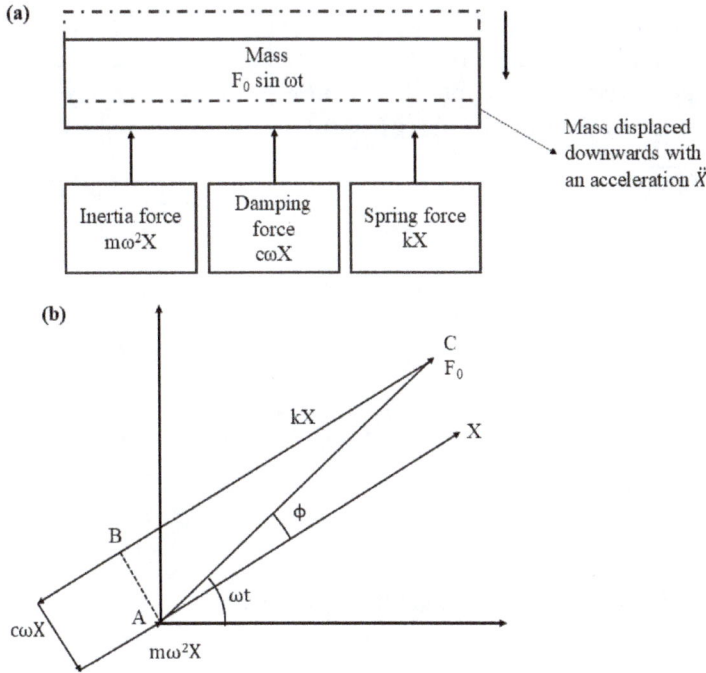

Figure 4.10 (a) Free body diagram of the different forces acting on the mass–vibration isolator–foundation system, (b) vector representation of forces acting on the mass–vibration isolator–foundation

The forces transmitted to the foundation are spring force and damping force. The free body diagram of the forces acting on foundation and their vector representation is shown in Figure 4.11. The elastic and damping force has a phase difference of $\pi/2$. The transmitted force is drawn and can be represented as the hypotenuse of the right-angled triangle EFG.

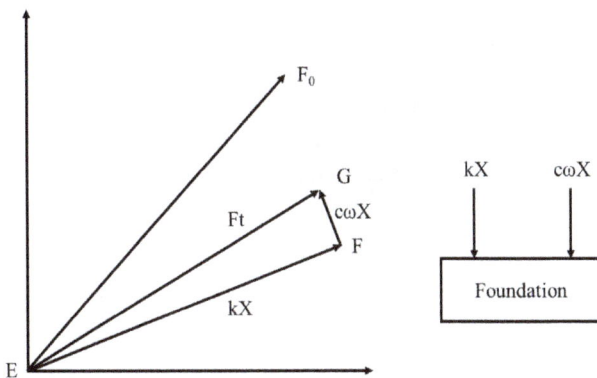

Figure 4.11 The different forces acting on the foundation vector diagram, b) free body diagram for the different forces acting on the foundation

From the vector representation of force transmission to the foundation, the transmitted force is the resultant of the spring force and the damping force (viscous force). These two forces act at a phase difference of $\pi/2$. Therefore, the transmitted force (Ft) is the resultant vector of the spring and damping forces:

$$Ft = \sqrt{(c\omega X)^2 + (kX)^2} \tag{4.39}$$

$$Ft = X\sqrt{(c\omega)^2 + (k)^2} \tag{4.40}$$

$$Ft = Xk\sqrt{\left(\frac{c\omega}{k}\right)^2 + 1} \tag{4.41}$$

Substituting for X:

$$Ft = \frac{F_0\sqrt{(c\omega)^2 + (k)^2}}{\sqrt{(k - m\omega^2)^2 + (c\omega)^2}} \tag{4.42}$$

Multiply and divide numerator by k and on simplifying, we get,

$$Ft = \frac{F_0 k\sqrt{\left(\frac{c\omega}{k}\right)^2 + 1}}{k\sqrt{\left(1 - \frac{m\omega^2}{k}\right)^2 + \left(\frac{c\omega}{k}\right)^2}} \tag{4.43}$$

$$Ft = \frac{F_0\sqrt{\left(\frac{c\omega}{k}\right)^2 + 1}}{\sqrt{\left(1 - \frac{m\omega^2}{k}\right)^2 + \left(\frac{c\omega}{k}\right)^2}} \tag{4.44}$$

Transmissibility (T_r) is:

$$Ft/F_0 = \frac{\sqrt{(c\omega)/k)^2 + 1}}{\sqrt{((1 - m\,\omega^2/k))^2 + ((c\omega)/k)^2}} \tag{4.45}$$

Generally, damping reduces transmissibility, because it lowers the amplitude of oscillation. The damping ratio is a key system parameter that measures how rapidly the oscillations decay from one bounce to another. It denoted by ζ and is defined as

$\zeta = c/2m\omega$

Where

m = mass

ω = natural frequency

c = damping coefficient

The ideal damping condition is known as critical damping, and represents minimum amount of damping required to prevent oscillation altogether.

The equation for transmissibility can be written in terms of damping ratio and frequency ratio by replacing the terms $c\omega/k$ and $m\omega^2/k$

Using Equation 4.6, Equation 4.17 and Equation 4.20

$$\frac{c\omega}{k} = \frac{c}{c_c} \times \frac{c_c}{2m} \times \frac{2m}{k} \times \omega = 2\zeta\omega/\omega_n$$

Using Equation 4.6

$$m\frac{\omega^2}{k} = (\omega/\omega_n)^2$$

Applying these to Equation 4.45

$$\frac{Ft}{F_0} = \frac{\sqrt{(2\zeta\omega/\omega_n)^2 + 1}}{\sqrt{(1 - (\omega/\omega_n)^2)^2 + (2\zeta\omega/\omega_n)^2}} \qquad (4.46)$$

If damping is negligible, Equation 4.46 can take the form

$$Tr = 1/(1 - (\omega/\omega_n))^2 \qquad (4.47)$$

Under conditions of resonance ($\omega = \omega_n$),

$$\text{Transmissibility} = \frac{\sqrt{1 + (2\zeta)^2}}{\sqrt{(2\zeta)^2}}$$

The damping capacity of rubber is very important during conditions of resonance.

Tr in terms of frequency ratio $r = \frac{\omega}{\omega_n}$

$$Tr = \frac{\sqrt{(2\zeta r)^2 + 1}}{\sqrt{(1 - r^2)^2 + (2\zeta r)^2}} \qquad (4.48)$$

$$\tan\varphi = c\omega X/kX - m\omega^2 X \qquad (4.49)$$

In terms of frequency ratio and damping ratio, $c\omega/k = 2\zeta\omega/\omega_n$, $m\omega^2/k = (\omega/\omega_n)^2$

$$\tan\varphi = \frac{2\zeta\omega/\omega_n}{1 - (\omega/\omega_n)^2} \qquad (4.50)$$

(φ is the phase lag between input excitation and spring force, α is the phase lag between transmitted force and spring force)

$$\tan\alpha = c\omega X/kX \qquad (4.51)$$

$$\tan\alpha = c\omega/k = 2\zeta\omega/\omega_n$$

To evaluate the effectiveness of vibration isolation, it is important to understand transmissibility as a function of the frequency ratio, as illustrated in Figure 4.12, which provides the following information:

At low frequency ratios (ω/ω_n = 0.1 to 0.3), the transmissibility T is close to 1, indicating that the amplitude of motion of the mass is almost the same as the amplitude of motion of the spring.

When the excitation frequency ω approaches the natural frequency ω_n, the system enters a condition of resonance, resulting in amplified motion of the mass.

The point at which ω/ω_n, = $\sqrt{2}$ is called the cross-over point. After this point, when the disturbing frequency exceeds the natural frequency, transmissibility is low and vibration isolation becomes effective.

Beyond the cross-over frequency, transmissibility decreases more for materials with a lower damping ratio while systems with higher damping exhibit reduced peak amplification.

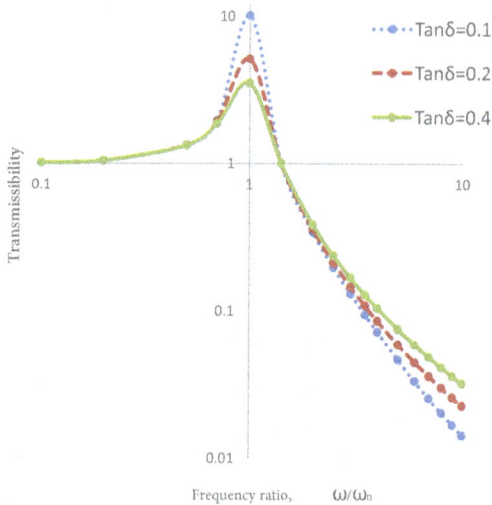

Figure 4.12 Graph of transmissibility versus frequency ratio for different values of tan δ (assuming $2\zeta = \tan\delta$)

From the graph, it is evident that there exists an amplification region within a specific range of frequency ratios and an isolation region at other ratios. A higher damping ratio improves the system's performance at any frequency. When the damping ratio is greater than 1, vibration transmissibility is less than one and isolation is good.

In designing an elastomeric vibration isolator, it is essential to know the disturbing frequency and the desired transmissibility level. The isolator is then designed with appropriate stiffness and natural frequency. The lower the isolator's natural frequency relative to the disturbing frequency, the greater is the vibration isolation efficiency. In such cases, low-damping elastomers are preferred.

However, when there are numerous disturbing frequencies, high-damping elastomers are preferred. At resonance, these are better at reducing impact forces than a more resilient rubber. Therefore, high hysteresis is required for effective impact absorption.

The energy loss in rubber is governed by its elastic and viscous components. Under sinusoidal deformation, the viscous component is out of phase with the elastic component.

Natural rubber and polychloroprene rubbers have very high resilience, while butyl rubber and nitrile rubber possess inherent damping properties. Generally, natural rubber is preferred for vibration isolation, due to its high resilience. A loss factor of 0.1 is generally considered the minimum for effective damping. While NR and CR may have low values of loss tangent depending on formulation, their damping behavior is close to that of EPDM.

Butyl rubber exhibits a loss tangent approximately four times greater than that of natural rubber (NR). This is attributable to its unique structure, featuring two geminal methyl groups on alternating carbon atoms. The resulting larger carbon–carbon bond angle allows tighter chain packing, reducing free volume relative to NR. This structural arrangement contributes to high damping characteristics, greater density and excellent impermeability. Additionally, butyl rubber offers high resistance to ozone and oxidation. The presence of methyl groups facilitates interactions between adjacent molecules, leading to a delayed response to deformation. It also offers a broad glass transition temperature.

Silicone rubber is suitable for both low- and high-temperature applications and is non-toxic. Its shock-absorbing properties make it widely used in applications such as engine mounts [2, 3].

Numerical example 1 (Force transmissibility and vibration isolation)

A piece of testing equipment with a mass of 800 kg rotates at 1800 rpm when a force of 2500 N is applied. A vibration isolator, in the form of a pad, is provided to protect the foundation. The pad becomes compressed by 1.5 mm, due to the weight of the mixing equipment. The isolator pad has a damping ratio of 0.2. Calculate the

(i) Amplitude of vibration of the mixing equipment

(ii) Force transmitted to the foundation

(iii) Transmissibility

Answer:

Mass m = 800 kg, F_0 = 2500 N Ω omega = 1800 x 2 x 3.14/60 rad/s = 188.4 rad/s

Static deflection δ = 1.5 x 10^{-3} m

$$K = mg/\delta$$

$$= 800 \times 9.81/1.5 \times 10^{-3} \text{m} = 5.23 \times 10^6 \text{N/m}$$

$$\omega n = \sqrt{k/m}$$

$$= \sqrt{\frac{5.23 \times 10^6}{800}} = 80.85 \text{ rad/s}$$

$$\frac{\omega}{\omega n} = \frac{188.4}{80.85} = 2.33$$

$$X = \frac{F_0/k}{\sqrt{(1 - (\omega/\omega_n)^2)^2 + (2\zeta \, \omega/\, \omega_n \,)^2}}$$

$$X = \frac{2500/(5.23 \times 10^6)}{\sqrt{(1 - (2.33)^2)^2 + ((2 \times 0.2 \times 2.33)^2}} = 1.056 \times 10^{-4}\text{m} = 0.1056 \text{ mm}$$

Transmissibility =

$$\frac{\sqrt{(2 \times 0.2 \times 2.33)^2 + 1}}{\sqrt{(1 - (2.33)^2)^2 + (2 \times 0.2 \times 2.33)^2}} = 0.302$$

Force transmitted =

$$Ft = F_0 \times \text{Transmissibility} = 2500 \times 0.302 = 755$$

Numerical example 2 (Motion transmissibility and vibration isolation)

A car with a mass of 1000 kg moves over a speed-breaker road surface that varies sinusoidally with an amplitude of 0.1 m and a wavelength of 6 m. The car's suspension system has a spring constant of 400 kN/m and a damping ratio of 0.5. If the car speed is 40 km/h, calculate the displacement amplitude of the car.

Answer:

Given m =1000 kg, λ wavelength = 6 m, amplitude Y = 0.1 m, **speed of car** = $\frac{40\text{km}}{h}$

Let X be the displacement amplitude of the car

$$\zeta = 0.5$$

For a car-disturbing frequency $\omega = 2\pi \, v/\lambda$, where V is the velocity in m/s, λ is the wavelength in m:

$$2\pi \, r = \lambda$$

And

$$v/r = \omega$$

Disturbing frequency, $\omega = v/r$, $2\pi\,r = 6$ m, r = 6 m/2π, r = $\frac{6}{2\times 3.14}$

$$\omega = \frac{40 \times 1000 \times 2 \times 3.14}{3600 \times 6} = 11.629 \text{ rad/s}$$

$$\omega_n = \sqrt{400 \times 1000/1000} = 20 \text{ rad/s}$$

$$\omega/\omega_n = 11.629/20 = 0.58$$

$$X/0.1 = \frac{\sqrt{(2 \times 0.5 \times 0.58)^2 + 1}}{\sqrt{(1 - (0.58)^2)^2 + (2 \times 0.5 \times 0.58)^2}} = 1.3$$

X = 1.3 × 0.1 = 0.13 m

That is, the 0.1 m bump on the speed-breaker road is transmitted as a 0.13 m bump to the chassis and the passengers of the car. The amplitude transmitted is higher because the frequency ratio is less than 1.

4.6 Compound Development for an Engine Mount Based on Natural Rubber

Engine mounts, as the name suggests, support the engine so that it remains securely in position and can function properly. During operation, an engine generates various movements and low-amplitude, high-frequency vibrations due to combustion events, such as the ignition of fuel by spark, and to interactions among moving parts all in the space of seconds. Vibrations and movements can be generated in different parts of a vehicle from the road surfaces over which the vehicle moves. Engine mounts are components that absorb these vibrations by different mechanisms involving energy storage or energy dissipation.

Engine mounts should have low modulus and high elasticity characteristics for high-frequency, low-amplitude vibrations, but more hysteresis and stiffness for high-amplitude vibrations. Natural rubber is highly useful for engine mounts. However, when oil resistance is required, neoprene rubber is preferred. If more vibration damping is required for a wide range of frequencies, the preferred rubber is silicone rubber. It has the added advantage of being non-toxic, has excellent vibration damping characteristics and is flexible at both low temperatures and high temperatures (–50 °C to +250 °C). For this reason, engine mounts and general vibration-absorbing mounts in submarines are made of silicone rubber. EPDM rubber can also be used as it has excellent heat resistance. The essential parts of an engine mount are rubber, metal plates and connection bolts. The metal plates are generally made of low-carbon steel or aluminum, on account their light weight and strength. Photographs of automotive engine mounts are shown in Figure 4.13.

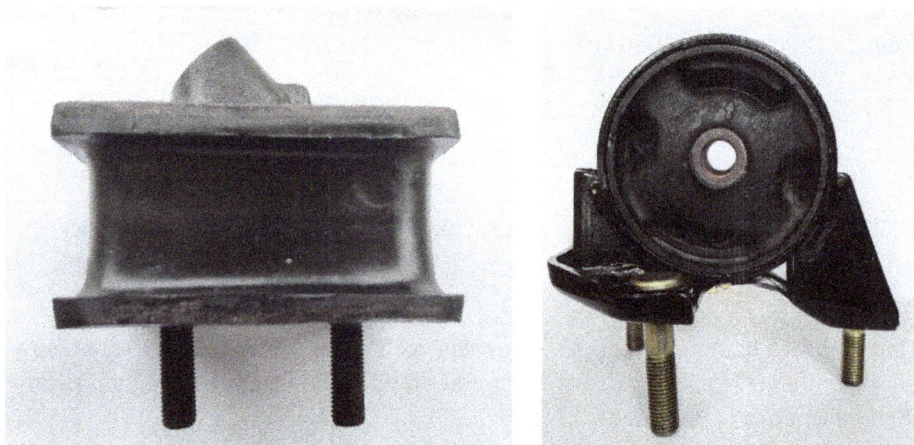

Figure 4.13 Engine rubber mounts with connection bolts (Courtesy: NIRT, Rubber Board, Kerala)

In general, the special features of engine mount compounds are high elasticity, low hysteresis and excellent dynamic properties, good strength combined with suitable stiffness characteristics that control the natural frequency of a rubber mount and related properties, such as low heat build-up good fatigue resistance, aging resistance and low compression set. It is known that the vulcanization system, type of accelerators chosen for cross-linking, the type and dose of anti-degradants and the type and quantity of fillers used affect the mechanical and dynamic properties. The inherent elastomeric properties can be retained to a great extent by using fillers with very low structures that do not enter into weak or chemical interactions with rubber molecules. These fillers maintain the high rebound, low hysteresis and low tan delta inherent to natural rubber.

The efficient vulcanization system containing CBS/TMTD has a synergistic effect on vulcanization characteristics. The aging properties and mechanical and dynamic properties at high temperatures remain good in the presence of suitable anti-oxidants and anti-ozonants.

Natural rubber owes its elasticity to its high molecular weight and long-chain branching. Sheet or premier quality block natural rubber can be used.

Table 4.1 presents a formulation for an engine mount based on natural rubber and thermal black, designed so as not to adversely affect resilience.

For superior dynamic properties in engine mounts, having the lowest possible values for dynamic stiffness and tan delta is desirable. Among the various carbon black grades, thermal blacks have the least adverse effect on dynamic properties.

The strength parameters can be increased with carbon black of smaller particle size and silica filler. If fillers are dispersed in the latex stage, improved filler dispersion can lead to lower heat build-up and reduced loss tangent. A carbon black and silica formulation with low heat build-up is presented in Table 4.2.

Table 4.1 Formulation for an Engine-Mount Rubber Compound

Ingredient	Phr
Natural rubber	100
Zinc oxide	4
Stearic acid	2
Thermal Black (N990)	40
SRF Black (N762)	20
Naphthenic oil	6
Antioxidant TMQ	1
Tetramethylthiuram disulphide	0.5
CBS	1.0
Sulfur	1.5

Table 4.2 Formulation for a Natural Rubber Compound Containing Carbon Black and Silica

Ingredients	Phr	
	MB45	MB 50
NR#	100	100
ZnO	5	5
Stearic acid	1	1
ISAF carbon black*	40	40
Ppt silica	5	10
TMQ	1	1
6PPD	1	1
CBS	1	1
S	2.5	2.5

* The calculated quantity of carbon black was added as a 20% dispersion, and the silica as a 25% dispersion in water to fresh latex of known dry rubber content and coagulated as carbon black rubber masterbatch to yield 45 or 50 phr carbon black/silica filler. No process oil was added during mixing.

Table 4.3 Mechanical Properties of Carbon Black and Silica-filled Rubber Compounds

Test Parameter	Compound ID	
	MB 45	MB50
Modulus 300 %, MPa	11	12
Tensile strength, MPa	28	28
Elongation at break, %	600	620
Tear strength, kN/m	102	113
Heat build-up, ΔT in °C (Goodrich flexometer)	16	18
DMA tan delta (dynamic strain: 0.0012, frequency: 10 Hz, temperature: 60 °C)	0.08	0.086

In anti-vibration rubber products, other synthetic rubbers employed are polychloroprene rubber, butyl rubber, hydrogenated nitrile rubber and silicone rubber. The strength characteristics of chloroprene and HNBR are close to those of natural rubber while HNBR, silicone and butyl rubber exhibit higher damping properties than CR or NR.

Silicone rubber stands out for its unique ability to withstand high and low temperatures. While natural rubber and polybutadiene rubbers perform well at low temperatures, they are unsuitable for continuous use above 70 °C. In contrast, fluorocarbon rubber is used for space applications, due to resistance to high temperatures, but is not ideal for low temperatures. Silicone rubber can be used continuously at high temperatures close to 150 °C with minimal change in its vibration isolation capabilities. Silicone rubber is bio-compatible and non-toxic. It has excellent resistance to aging.

4.7 Shock Isolation

Shock happens in a body when the high kinetic energy of a mass is suddenly transferred to a system. It is characterized by a rapid change in acceleration and is often observed in everyday situations. While vibration is a periodic motion, shock is a non-periodic transient event. A simple definition of shock is the transmission of kinetic energy to a system within a short timeframe, typically much shorter than that of the natural period of vibration of the system. The effect of an impulse force is perceived as shock in the system. In a collision of two objects, an impact force arises as a result of a sudden change in velocity, leading to the transfer of kinetic energy between the objects.

Vehicles driving over bumps or gutters on roads, applying sudden brakes in cars, striking an object with a hammer, a heavy metal ball falling from a height are further examples of shock. Many electronic machines are subjected to shock when in service or being transported on high-speed bumpy roads. In the case of a rough road, the input shock may be displacement, but the effect or response will take the form of acceleration.

Rubber pads used in vehicles, mats used on floors and elastomer bearings used in bridges are materials that act as shock absorbers. Rubber shock isolators can absorb part of the kinetic energy from shock, dissipate it as heat, store some of the kinetic energy and then release it at a slow rate.

4.7.1 Shock Pulse

Shock is a sudden transient force that disturbs the equilibrium of a system and results in a rapid change in velocity. This abrupt change in velocity or momentum is known as shock excitation. Shocks can be categorized based on the nature of the excitation as impulse force shock or velocity shock.

Though shocks can be categorized based on the nature of excitation as velocity shock and impulse force shock, in reality in both cases the change in a body's velocity happens due to the impulse force which is related to the body's momentum. So, velocity and momentum are involved in both impulse force shock and velocity shock. In both cases high impulse of energy is transmitted resulting in increase in body's acceleration in a short time.

If the shock excitation is due to an impulse force, the shock is called impulse force shock. Here, the force is equivalent to the magnitude of change in momentum of the body of mass m, leading to an instantaneous change in velocity. Examples are the force exerted on a ball struck by a bat and the force on a nail hit by a hammer.

If the shock is caused by a sudden change in velocity from an initial velocity ($V1$) to another velocity ($V2$), it is called velocity shock. A free-fall impact also imposes velocity shock on a system. During sudden velocity change, as in velocity shock, the mass experiences an instantaneous change in velocity. An example of velocity shock is a bullet impacting a target, events like explosions that cause very sudden change in a body's velocity. For an impact with no rebound (inelastic impact), the velocity change Vi is given by:

$$Vi = \sqrt{2gh} \tag{4.52}$$

For impact with complete rebound (elastic impact), the velocity change Ve is given by:

$$Ve = 2\sqrt{2gh} \tag{4.53}$$

Where

g = acceleration due to gravity

h = drop height

Shock is graphically expressed as a shock pulse, a plot of acceleration as a function of time. These plots provide information about shock duration, maximum acceleration amplitude and the pattern or shape of the pulse. Shock pulses (Figure 4.14 a, b, c) can take on various forms, such as half-sine, triangular and rectangular [4, 5].

4.7.2 Mathematical Expression for Velocity Changes Occurring During a Shock

Mathematically, shock is represented by a change in velocity. The severity of a shock is assessed by the velocity change that happens during the shock excitation.

The half-sine, triangular and rectangular pulses shown in Figure 4.14 (a, b and c) each exhibit a peak acceleration amplitude $A0$ and a time duration $\tau0$. When these pulses are idealized as sinusoidal, it is possible to determine the change in velocity by integrating the area under the curve. A velocity shock showing change in velocity is illustrated in Figure 4.14 d. Velocity shock represents a sudden change in velocity over a certain time duration.

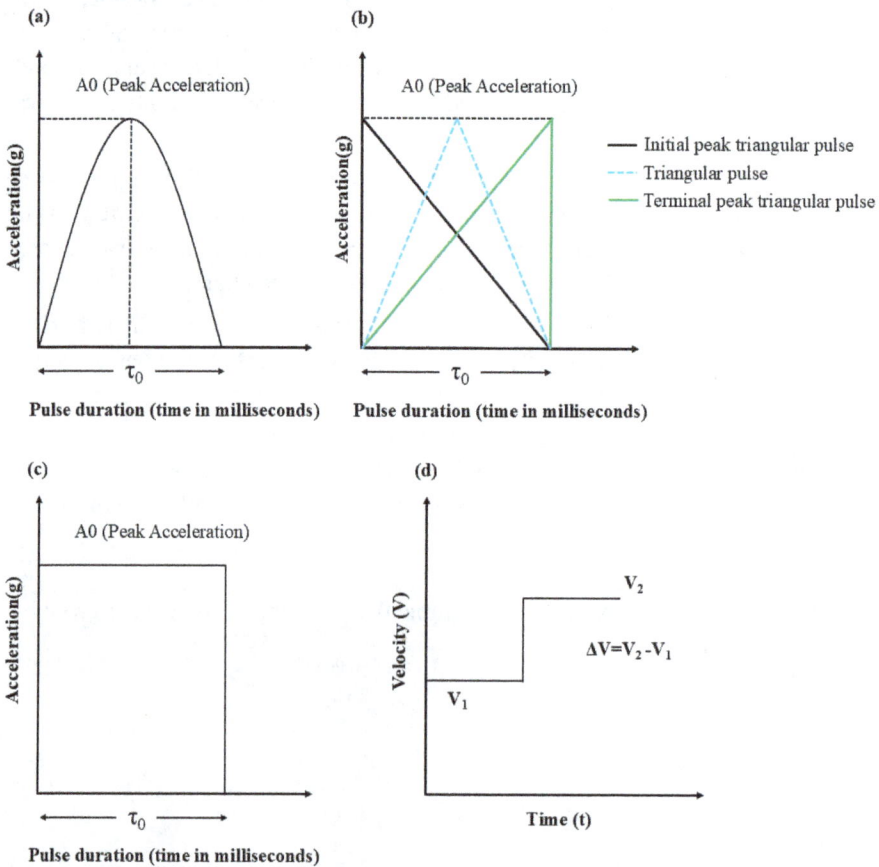

Figure 4.14 Shock pulses of different shapes: (a) half sine, (b) triangular, (c) rectangular, and (d) velocity shock

As a general form, the motion of a shock pulse can be modeled with a sine curve, where the acceleration at any time $A(t)$ is expressed as:

$$A(t) = A0 \sin 2\pi f t \tag{4.54}$$

Here, A0 is the peak acceleration. Since the shock pulse is half of one period of a sine wave, the frequency f associated with this duration ($\tau 0$) is $f = 1/2\tau_0$.

Substituting into the acceleration equation:

$$A(t) = A0 \sin(\pi/\tau 0)t \tag{4.55}$$

The area under the half cycle of the acceleration graph over the time interval, τ_0 is equal to the change in velocity V during that time interval. The velocity change is based on the area of a sine curve:

$$
\begin{aligned}
V &= \int_0^{\tau 0} (A0 \sin(2\pi f t)dt \\
&= \frac{-A0}{2\pi f}(\cos \pi - 1) \\
&= \frac{A0}{\pi f}(\text{as } \cos \pi = -1)
\end{aligned}
$$

Substituting $f = 1/2\ \tau 0$

$$V = A0/\pi f \tag{4.56}$$

$$V = A0/\pi(1/2\tau 0) \tag{4.57}$$

$$V = 2\ A0\ \tau 0/\pi \tag{4.58}$$

The half-sine shock pulse shown is a plot of acceleration amplitude expressed in terms of acceleration due to gravity g ($9.81\ \text{m/s}^2$) (Figure 4.10 a).

For practical applications, the equation is written with acceleration expressed in multiples of g:

$$V = 2g\ A0\ \tau 0/\pi \tag{4.59}$$

The velocity change is based on area of square and triangular gular shock pulse and change in velocity is as follows:

For a square shock pulse:

$$V = gA0\ \tau 0 \tag{4.60}$$

For triangular shock pulse:

$$V = \frac{1}{2}\ gA0\tau 0 \tag{4.61}$$

For a velocity shock change in velocity $\Delta V = V_2 - V_1$

Shock input is based on velocity change rather than acceleration change and shock input is based on these equations.

4.8 Shock Isolation

The effect of shock in a system can be reduced by inserting a shock absorber with appropriate stiffness and damping characteristics. In shock isolation, the goal is to reduce the amplitude of the resonance condition. The roles and effects of stiffness and damping in isolating shock differ from those in steady-state vibration isolation.

During a shock event, energy is transferred rapidly to a system and with large acceleration. Shock isolators are employed to protect surrounding structures or sensitive components by limiting and absorbing these forces. The isolators store the shock energy and then release it over a longer period corresponding to the system's natural frequency (Figure 4.10 a and b). Energy storage takes place through deflection of the isolator. The release of the strain energy stored in the isolator causes the isolated body to vibrate at the natural frequency until the damping mechanism has dissipated the energy. The effectiveness of a shock isolator is measured as the ratio of dynamic output to dynamic input based on acceleration and factors like ability to absorb the kinetic energy and to undergo deflection.

4.8.1 Mathematical Expression for Linear Deflection of an Isolator under Shock

The impact of a force can be observed a s a static deflection of the isolator. The isolator partially stores the shock pulse's kinetic energy, which is related to the static deflection, say X. The work done by the isolator is equivalent to the elastic potential energy or strain energy in the isolator. If k is the spring stiffness, X is the deflection, m is the mass of the object that impacts or falls with a velocity V, then:

$$^1/_2 kX^2 = {}^1/_2 mV^2 \tag{4.62}$$

$$kX^2/m = V^2 \tag{4.63}$$

$$\sqrt{k/m}\, X = V \tag{4.64}$$

$$X\, 2\pi f0 = V \tag{4.65}$$

Or

$$X = V/2\pi f 0$$

Consider a spring that is used as a shock isolator with a natural frequency of $f0$. In this case, the spring will move a distance X, called deflection, when subjected to a static or dynamic force, F. There can be deflection in static and dynamic conditions. In the design of shock absorbers, it is important to know the dynamic deflection, the spring force of the isolator and the velocity change of the shock excitation.

4.8.2 Mathematical Expression for Shock Output (G_{out})

Harmonic oscillation (Figure 4.2, spring-mass system) can be represented as the displacement of the block as a function of time, t. Based on Hooke's law and Newton's law, the differential equation for the motion of a spring mass system is:

$$m\ddot{X} + c\dot{X} + kX = 0$$

A general solution to this differential equation for the displacement of mass as a function of time is:

$$X(t) = Xe^{i\omega t} \tag{4.66}$$

$$\dot{X}(t) = i\omega Xe^{i\omega t} \tag{4.67}$$

$$\ddot{X} = \omega^2 Xe^{i\omega t} \tag{4.68}$$

$$\ddot{X} = \omega^2 X(t)$$

$$\ddot{X} = i\omega i\omega Xe^{i\omega t}$$

$$\ddot{X} = i\omega \dot{X}$$

The i term indicates phase. The magnitude relationship between acceleration and velocity is given by:

$$|\ddot{X}| = |\omega \dot{X}(t)| \tag{4.69}$$

$$\ddot{X} = 2\pi f \dot{X}(t) \tag{4.70}$$

The shock output is expressed in terms of acceleration and is related to the velocity change (V) [6].

Shock output designated as G_{out} in multiples of g is expressed as:

$$G_{out} = 2\pi f nV \tag{4.71}$$

4.8.3 Shock Transmissibility

The shock acceleration level experienced by a piece of equipment is commonly expressed in terms of 'g- level', a dimensionless term representing multiples of acceleration due to gravity g. This term is used specifically to describe the shock which equipment can withstand without failure. The maximum g-level that equipment can endure without failure is referred to as its fragility. The degree of protection from shock is a function of the fragility [7].

Vibration isolators are designed for maximum allowable deflection, maximum allowable transmitted force and acceptable velocity change.

One of the key response parameters of a system (foundation) protected by a shock isolator is transmissibility (Tr).

The input acceleration from a shock pulse can be expressed in terms of velocity change, as shown in Equation 4.59. For a half-sine shock pulse, the shock input G_{in} is:

$$G_{in} = 2gA0\ \tau o/\pi \tag{4.72}$$

The shock output G_{out}, related to the system's response, is given by:

$$G_{out} = 2\pi f nV/g$$

Shock transmissibility is defined as:

$$Tr = \frac{\text{shock output } (G_{out})}{\text{shock input } (G_{in})} \tag{4.73}$$

$$Tr = \frac{\text{output acceleration}}{\text{input acceleration}} \tag{4.74}$$

The rubber shock isolator absorbs or stores kinetic energy from the shock. The output shock (G_{out}) is related to the kinetic energy stored and released by the shock absorber over a longer time (Figure 4.15).

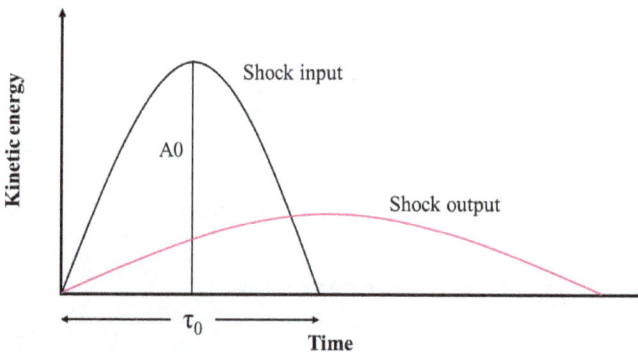

Figure 4.15 The kinetic energy of a half-sine shock pulse released over a longer period of time through the use of a shock isolator

The output of shock response can be represented as a static deflection X of the rubber shock absorber, generated by the shock force (as defined in Equation 4.65). This deflection initiates vibration of the shock absorber at the system's natural frequency, such that the shock is transmitted over a longer period of time than original shock pulse (Figure 4.16).

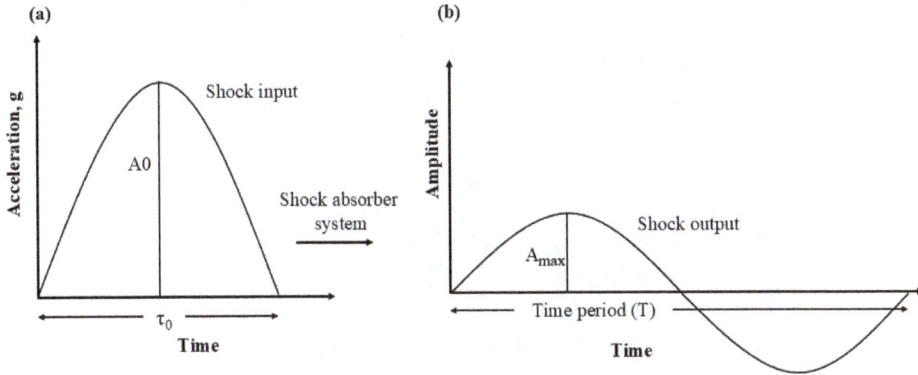

Figure 4.16 Half-sine shock pulse, (b) dissipation of absorbed shock as vibration of shock absorber at its natural frequency with time period T

The shock output is thus related to the velocity change (V) of the shock pulse and the natural frequency of the shock absorber system (f0) affected by the shock.

Numerical question for designing a shock isolator

An electronic machine weighing 5 pounds is subjected to a 12-millisecond, half-sine pulse with an amplitude of 18 g and the electronic machine needs to be isolated from this shock by a shock isolator mount. The fragility of the equipment, which is the maximum input shock acceleration, is rated at 12 g. Calculate the natural frequency of the shock isolator mount and the dynamic deflection.

Answer:

The input shock can be quantified by the velocity change as per the equation given below:

$$V = 2g \, A0\tau0/\pi$$

$$V = 386 \text{ inch/s}^2$$

$$A0 = 18 \text{ g}$$

$$\tau0 = 12 \text{ ms} = 0.012 \text{ s}$$

$$V = 2 \times 386 \text{ inch/s}^2 \times 18 \text{ g} \times 0.012 \text{ s}/3.14 = 53.11 \text{ inch/s}$$

The allowable shock based on fragility is then (or G_{out} allowed is 12 g or a force corresponding to 12 g shock is allowable and can be accommodated by the shock absorber):

$$G_{out} = 2\pi f0 V/g$$

$$f0 = G_{out} \times g/2 \times 3.14 \times 53.11 \text{ inch/s}$$

$$f0 = 12 \times 386 / 2 \times 3.14 \times 53.11 = 13.89$$

Or

$$f0 = 14 \text{ Hz}$$

Dynamic deflection $(X) = \frac{V}{2\pi f0}$
$$= \frac{53.11}{2 \times 3.14 \times 14} = 0.604 \text{ inches}$$

4.9 Rubber as a Sound-Absorbing Material

Sound is a form of energy produced when an object is disturbed and starts vibrating. It is produced by any material that has elasticity and mass. Just as a vibrating body has a frequency, the sound waves arising from vibration are also characterized by frequency. The combined study of sound and vibrations is referred to as acoustics. When transmitted through denser (compared to air) materials, such as wood, the sound energy manifests as vibration rather than audible sound, due to its very low frequency. We hear sound only if its waves have a frequency of more than 20 Hz. The audible frequency range for humans is 20 to 20,000 Hz. Sound waves with a higher frequency than 20,000 Hz are called ultrasonic waves. As the frequency of sound increases from 20 Hz, it becomes shrill or higher in pitch. Similarly, greater amplitude or a larger vibrating surface leads to increased loudness.

Sound is perceived as noise and music. Music is perceived as pleasant, but noise is unwanted and typically unpleasant, being characterized by a highly variable pitch and lacking any meaningful communication. Loud noises from a crowd, noise produced by operating machinery, moving traffic and natural calamities can be annoying and unwanted sounds pollute the environment. Prolonged exposure to loud noise also adversely affects health. Hence, noise needs to be reduced. Additionally, reflection of sound creates echoes, which are also undesirable. So, there is a need to mitigate sound. Sound absorbers can mitigate unwanted noise.

Noise and vibration can be reduced with vibration isolators, sound absorbers or acoustic enclosures. Sound-absorbing materials can be of low weight and hence used for effective sound absorption in the construction of buildings, vehicles, trains, aircraft, ships and space shuttles requiring sound and vibration reduction or isolation. Fabric-wrapped panels can be good acoustic absorbers. Rubberized asphalt and stone can reduce traffic noise.

4.9.1 Propagation of Sound in a Medium

Sound propagation needs a medium (water, air) because vibration produced by a source causes the medium around it to vibrate. In air, the energy due to vibration compresses some of the surrounding air molecules, which compresses the adjacent air molecules, causing a periodic pressure difference. The areas of low and high pressure are called compressions and rarefactions. Sound energy is transmitted as a wave when particles of a medium vibrate and touch other particles of the medium. They travel as longitudinal waves. Sound is propagated or transmitted as waves through solids, liquids and gases. Because it needs a medium for propagation, sound does not get transmitted through a vacuum.

When sound encounters a solid medium, some is reflected back, some is absorbed and some is transmitted to another medium, depending on the wavelength of sound. Soft polymeric materials such as fibers absorb sound. Hard surfaces such as ceramic tiles and wooden planks reflect sound waves, causing echoes. Substances such as concrete and brick transmit sound vibrations of low frequency. Sound absorption is defined as the amount of energy removed from the sound wave as the wave passes through a given thickness of material.

4.9.2 Key Terms Related to Sound Absorption

4.9.2.1 Sound Pressure

Due to the energy of sound, the pressure of the medium changes instantaneously at that point in the medium where there is a sound wave. Sound pressure is the difference between the instantaneous pressure at a point in the presence of a sound wave and the static pressure of the medium. Sound pressure indicates the amplitude of the sound at a specific location in space and is a scalar quantity. Just as a temperature variation is observed in different parts of a room where a heater is placed, there is a variation of sound pressure at different locations in a room with a source of sound. The unit of sound pressure is Pascal. A loud noise usually corresponds to a large pressure variation and a weak one produces only minor variation. The softest sound an average human ear can detect has a pressure variation of 20 microPascals; this is taken as zero decibels (dB). The human ear can hear sounds up to 130 dB.

4.9.2.2 Sound Power

This is the rate at which sound energy is emitted per unit time. Sound power is also measured in Watts. (A heater that produces a particular amount of heat per hour has units of watts).

4.9.2.3 Sound Intensity

The intensity of a sound wave is defines as sound power per unit area. It is the average amount of energy transmitted per unit time through a unit area in a specified direction.

4.9.3 Sound Absorbers

As the name suggests, sound-absorbing materials absorb, reflect or transmit minimal sound. Synthetic fibers and other polymers have been used as sound-absorbing materials. A material's sound-absorbing property depends on the sound frequency, the thickness of the material, surface finish and chemical composition. Sound absorption quality increases when the materials are made porous. Polymers are lightweight materials that are readily processable and can easily be converted into porous forms such as sponge or foam. Elastomers and amorphous polymers typically exhibit better sound absorption properties than semicrystalline or crystalline materials.

Undesired sound can be absorbed by soft and porous substances, cavity resonators and non-perforated membranes, collectively referred to as acoustic protection. Porous absorbers have a network of minute pores within a soft, solid frame. Textiles, carpets, foam and sponges are examples of porous materials. Enclosed bodies of air within rigid walls connected by narrow openings are called cavity resonators. Mineral wool enclosed in semi-rigid containers functions as a non-perforated membrane and has sound-absorption qualities. Sound can be reduced by blocking transmission. An example is a double-walled acoustic panel. Smooth and rigid surfaces reflect sound, while soft and irregular surfaces absorb it.

4.9.4 Sound Absorption Characteristics

Sound is "absorbed" when sound energy is converted into heat within the material, thereby reducing sound pressure. In polymers, sound absorption primarily occurs through the vibration of long molecules, a phenomenon generally called damping or through friction between fluid molecules within the solid material. Sound absorption property is quantified by the sound absorption coefficient α, which is the ratio of absorbed sound energy to the incident sound energy. The two parameters of a material that affect sound absorption are porosity (air present in the material) and structural factors such as molecular weight, orientation of molecules etc.

4.9.5 Sound Absorption by Porous Materials

Porous materials are typically categorized as cellular (for example, sponge rubber), fibrous (cotton) or granular (rubber crumb). The porous materials absorb sound through frictional losses, momentum loss and temperature fluctuation. When sound energy meets a porous surface, as a general case, some sound energy is absorbed while a part of sound energy is reflected. The reflected wave encounters another cellular structure and the same absorption and reflection process continues.

A porous material such as a sponge is a solid matrix with many inter-connected minute pores and contains a solid and fluid phase. Sound waves can penetrate easily into these pores. Inside the pores, the waves do not have a continuous medium to travel through, but rather follow a tortuous path such that much of the energy is dissipated inside the material mainly through viscous shear and frictional loss before it is transmitted to the outside. The incident sound waves propagate through the air through longitudinal vibrations of air molecules. When they contact the pore boundary, they experience viscous shear (drag), leading to energy loss. It is also possible that the sound waves undergo multiple reflections from the solid surfaces inside the pores. Sound energy is lost as frictional losses at the air-solid interface occur during this time. The sound waves also penetrate inside the solid phase and the air vibrations can initiate the vibration of rubber molecules. The work involved in this process leads to sound energy dissipation, due to the hysteresis of the rubber.

The efficiency of sound absorption depends on both porosity and pore size. If the pore size is relatively small, then energy dissipation from the sound waves is relatively high, but a limited proportion of sound waves can penetrate the porous structure. However, if the pore size is relatively large, more sound energy can penetrate the porous structure, but energy dissipation from the sound waves is relatively low. Closed-cell foams make ineffective sound absorbers and the foam has to be open-celled in order to be a sound absorber.

The advantage of rubber is its viscoelastic nature. It has the ability to store the energy of a disturbance or deformation as strain energy and also the ability to dissipate energy by means of hysteresis. It can also easily be given different shapes by compression molding, calendering and extrusion techniques. Expanded rubber with interconnected cells can be prepared from latex and dry rubber in the conventional way.

4.9.5.1 Cellular Rubber-Based Acoustic Materials

Rubber that contains a mass of cellular structure is known as cellular rubber. Three types of cellular materials can be made from natural rubber: foam, sponge and expanded rubber. Foam is prepared from latex and the cells are intercommunicating. Sponges prepared from dry rubber have interconnecting cells. Generally, the cells of expanded rubber obtained from dry rubber are closed and not open. For sound absorption, open cells are preferred. For expanded rubber and sponges, chemicals called blowing agents are added during the compounding process. During vulcanization, the blowing agent releases gases that create a cellular structure in the rubber. The blowing agents benzene sulfonyl hydrazide (BSH) and Dinitroso Pentamethylene Tetramine (DNPT) liberate mainly nitrogen, while sodium bicarbonate produces carbon dioxide. The latter produces intercommunicating cells as carbon dioxide diffuses at a faster rate than nitrogen in rubber. So, nitrogen-liberating blowing agents produce closed cells, while carbon dioxide-liberating blowing agents produce open cells. Latex foam and sponge are cellular rubbers with interconnected cells.

4.9.5.2 Latex Foam

In a typical process (Dunlop method), the three major steps are foaming of the compounded latex, gelling of the foamed latex and vulcanizing of the gelled rubber foam. Air is introduced into natural rubber latex compounded with stabilizer, accelerator, activator and sulfur as vulcanizing agent. The level of foaming is attained by adjusting the dose and type of fatty acid soaps used. A higher quantity of fatty acid soaps and a blend of fatty acid soaps are used to ensure a good level of foaming during the mixing of latex compounds using a planetary mixer. Once the latex has expanded to a sufficient level, a cationic surfactant (acting as a secondary gelling agent) is introduced. Gelling agents such as zinc oxide and sodium silicofluoride are then added. The sensitized, expanded latex – still in its liquid phase – is subsequently transferred into suitable molds.

The steps in the preparation of latex foam are as follows:

First-stage compounding: Latex is compounded in the first stage with sulfur, accelerators and anti-oxidants as dispersion or emulsion. The quantity of fatty acid soap added is 0.2 phr.

Maturation: The compounded latex is kept for 16 h at room temperature to allow mixing and diffusion of compounding ingredients to the rubber particles.

Second-stage compounding: Matured latex is mixed with an additional quantity (about 1 phr) of fatty acid soap. Air is introduced into the latex compound using a Hobart planetary mixer run at a high speed of about 150 rpm by frothing action of the soap to the required level of expansion (about 7 times the original volume). The speed is then reduced to comminute bigger air bubbles to achieve a refined cell structure and then foam stabilizer is introduced. This is followed by addition of zinc oxide and sodium silicofluoride. The mixture is then transferred to previously heated molds, which are treated with a suitable mold-releasing agent. Within a few minutes, latex is converted into a solid gel. During this gelation process, the pH of latex decreases and the unconnected air bubbles break to form an interconnecting pore structure.

Vulcanization and removal from mold: The mold is placed in a steam chamber at 120 °C for about 45 minutes. During the vulcanizing or curing process, the material inside the mold is converted into latex foam. The latex foam is removed from the mold.

Gelation of latex occurs as a result of (a) the fall in pH due to the formation of HF, (b) the adsorptive effect of silicic acid and (c) the removal of adsorbed soap from the surface of rubber particles and their conversion into insoluble fatty acid soaps.

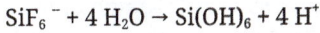

$Na_2SiF_6 \rightarrow 2\ Na^+ + SiF_6$

$SiF_6^- + 4\ H_2O \rightarrow Si(OH)_6 + 4\ H^+$

Apart from the destabilizing action of hydrofluoric acid, the powerful adsorptive effect of the silicic acid on stabilizers of the colloidal system plays a part in gelation. When ZnO is present together with soap, it also destabilizes latex by forming a zinc–amine complex. The complex dissociates to give Zn^{++} ions, which react with soap, causing destabilization by the formation of insoluble soaps.

Table 4.4 Typical Formulation for a Latex Foam

Ingredient	Dry Weight
Stage 1	
60% centrifuged latex	100
Potassium oleate 20%	0.2
Sulfur, 50% dispersion	2.0
Zinc diethyl dithiocarbamate, 50 % dispersion	1.0
Zinc mercapto benzothiazole, 50 % dispersion	1.0
Stage 2	
Potassium oleate, 20% solution	1.0
Cetyl trimethylammonium bromide, 30% solution	1.0
Zinc oxide, 50% dispersion	6
Sodium silicofluoride, 20% solution	2

4.9.5.3 Sponge Rubber

Polymers with fine cells, such as expanded rubber, are highly effective at absorbing low-frequency and high-frequency sound. Polyurethane (PU) foam and polyvinyl chloride (PVC) foam are the two most common materials that can be used for indoor sound absorption and noise reduction. However, these two types of foam materials inevitably pose environmental concerns during production and consequently have an effect on human health. Among various rubber types, silicone rubber is considered ideal from an environmental point of view.

Rubber sponge is prepared using CO_2-liberating blowing agents such as bicarbonates, alone or in combination with other nitrogen-gas-liberating blowing agents. In cellular rubbers or expanded rubber, properties can be varied not only *via* the type of compounding ingredients in the formulation, but also *via* the type and number of cells, to yield a versatile range of properties.

Both sponge and expanded rubber are conveniently made with blowing agents. A blowing agent should generate sufficient quantities of gases, such as CO_2, ammonia or nitrogen, at vulcanizing temperatures. Additionally, it should be non-toxic, colorless after processing and cost-effective.

The thermal decomposition of ammonium carbonate ($(NH_4)_2CO_3$) and ammonium bicarbonate (NH_4HCO_3) begins at 30 °C and 60 °C, respectively. Both blowing agents release ammonia gas and carbon dioxide gas.

$(NH_4)_2CO_3 \rightarrow 2\ NH_3 + CO_2 + H_2O$

$NH_4HCO_3 \rightarrow NH_3 + CO_2 + H_2O$

The blowing agent $NaHCO_3$ produces CO_2 gas upon decomposition:

$NaHCO_3 \rightarrow Na_2CO_3 + CO_2 + H_2O$

Open-cell rubber is generally produced by selecting a blowing agent with low temperature or rapid decomposition rate, such as sodium bicarbonate, p-toluene sulfonyl hydrazide (TSH) or 4,4′-oxybis-benzenesulfonylhydrazide (OBSH). Any combination of these will typically produce an open-cell sponge with its characteristic interconnected cells.

Table 4.5 Typical Formulation for Rubber Sponge

Ingredients	Phr
NR	100
ZnO	5
St acid	3
Antioxidant	1
China clay	30
Calcium carbonate	20
General purpose furnace black	10
Naphthenic oil	30
$NaHCO_3$	10
MBTS	1.5
TMTD	1.0
S	2.5

4.9.5.4 Polyurethane Foams

Polyurethane (PU) foams are widely used in various industries, particularly automotive applications, due to their lightweight, ease of manufacture and excellent sound absorption properties. PU foams are a class of versatile materials that are synthesized through the reaction of polyols and isocyanates [8, 9, 10]. The properties of polyurethane foams can be adjusted by varying the raw materials and incorporating different additives and nanomaterials [10]. PU foams are commonly used in automotive applications [9], for cushioning, molded flexible foams and interior trim components.

The production process of polyurethane foam involves mixing polyols, isocyanates, catalysts and other additives, followed by a foaming and curing stage. Polyurethanes are formed by the reaction of the hydroxyl group of the polyol with the isocyanate group, which releases carbon dioxide and causes the material to expand and solidify [10, 11].

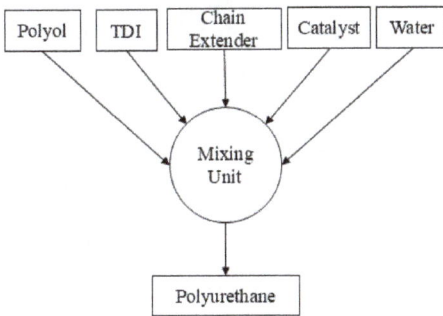

Figure 4.17 Diagram showing the production process for polyurethane

The reaction is carried out in the presence of a catalyst, such as stannous octoate and triethylenediamine [12]. A diagrammatic representation is shown in Figure 4.17. Industrially, PUs are produced by a batch process or a continuous slabstock process. A typical formulation for the production of PU foam with approximate density of 25 kg/m^3 is given in Table 4.6.

Table 4.6 A Typical Formulation for PU Foam

Sl. No	Ingredients	Parts
1	Polyol	100
2	Toluene diisocynate	65
3	Stannous octoate (catalyst)	0.2
4	Amine (catalyst)	0.28
6	Polysiloxane (surfactant)	1.9
7	Filler (CaCO$_3$)	20–40
8	Water	4.5

One of the key properties of polyurethane foams is their excellent sound absorption characteristics, making them suitable for use in noise-damping applications, such as theaters, offices, recording studios and other industrial requirements [11, 12]. Sound absorption properties are influenced by the cell structure, density and pore size of the foam, which can be tailored through the selection of raw materials and the processing conditions. When sound waves hit the surface of the open-cell foam, air passes through the material, causing the cell walls to stretch, bend and buckle [11, 13].

Research studies have reported that incorporating inorganic fillers into polyurethane foams can enhance sound absorption efficiency significantly, as these fillers absorb sound energy as it passes through the foam [13, 14]. Adding fillers such as silicone-acrylic particles, $CaCO_3$, magnesium hydroxide and wood fibers to PU foams can greatly enhance sound absorption by altering the cell structure and increasing the number of partially open pores [15, 16, 17, 18]. Graphene oxide-polyurethane aerogels and carbon nanotube-enhanced PU foams, as hybrid materials, offer improved sound absorption. Their unique microstructures allow for effective sound wave damping and scattering, improving overall sound absorption [19, 20].

Recent developments in polyurethane foams have focused on improving their sustainability, such as the use of renewable raw materials such as crude glycerol and all-water foaming techniques to reduce the impact on the environment. Key approaches in this area include optimizing foam morphology, integrating bio-based and recyclable materials, adding natural and synthetic fibers and recovering waste materials. These developments collectively contribute to more effective and environmentally friendly sound insulation solutions.

Appendix

1. Damping based on logarithmic decrease in amplitude

We can also consider damped oscillation as a decrease in height of or the amplitude of the oscillation. The amplitude decreases in successive cycles and this is called logarithmic decrement, expressed as

$$\text{Logarithmic decrement } \delta = \frac{\ln xi}{\ln xi + 1}$$

(If amplitude is xi and after one cycle there is a 5% reduction in amplitude, then the amplitude after one cycle is (1-.05) of xi = 0.95 xi, so δ = ln xi/0.95 xi = ln 1.5 = 0.048)

The damping factor ζ is related to the logarithmic decrement.

2. Mathematical expression for energy loss in cyclic deformation of rubber based on the area of the hysteresis loop

In sinusoidal rubber deformation with an angular frequency ω, stress lags behind the strain by an angle delta as given below. From Equations 3.50 and 3.51:

$$e(t) = e0 \sin(\omega t)$$

Where

$e0$ = maximum strain amplitude, ω the angular frequency (2π times the frequency in Hertz)

$e(t)$ = strain at any time t:

$$\sigma(t) = \sigma 0 \sin(\omega t + \delta)$$

The area of the hysteresis loop gives the energy lost during one deformation cycle. The area is obtained by integration of change in strain with respect to time over a time duration corresponding to a time period of one cycle, assumed to be $T = 2\pi/\omega$:

$$W2 = \int_0^T \sigma de$$

$$W2 = \int_0^T \sigma \left(\frac{de}{dt}\right) dt$$

Using

$$\sin(A + B) = \sin A \cos B + \cos A \sin B$$

Therefore

$$\sin(\omega t + \delta) = \sin \omega t \cos \delta + \sin \delta \cos \omega t$$

$$W2 = \omega\sigma 0 e 0 \int_0^T \cos \omega t [\sin \omega t \cos \delta + \sin \delta \cos \omega t] dt$$

$$= \omega\sigma 0 e 0 \int_0^T \cos \omega t \sin \omega t) \cos \delta \, dt + \int_0^T \cos^2 \omega t \sin \delta dt$$

$\cos \delta$ can be considered constant, so the integral of $\int_0^T \cos\omega t \sin \omega t$ will be $\frac{\sin^2 \omega t}{2} + C$ for $T = 2\pi/\omega$ Since $\sin 2\pi = 0$, the integral is zero:

$$2 \cos^2 \theta = \cos 2\theta + 1$$

$$\cos^2 \theta = \frac{1}{2} \cos 2\theta + \frac{1}{2}$$

$$\cos^2 \omega t = \frac{1}{2}\cos 2\omega t + \frac{1}{2}$$

$$= \omega\sigma0\epsilon0 \left[0 + \int_0^T \cos^2 \omega t \sin \delta \right] dt$$

$$= \omega\sigma0\epsilon0 \left[0 + \int_0^T \frac{1}{2}\cos 2\omega t + \frac{1}{2}\sin \delta \right] dt$$

The integral of first term is zero

$$= \omega\sigma0\epsilon0 \sin \delta \frac{1}{2} \int_0^T dt = \omega\sigma0\epsilon0 \sin \delta \frac{1}{2}T = \omega\sigma0\epsilon0 \sin \delta \frac{1}{2}2\pi/\omega = \epsilon0 \sin \delta = E''$$

$$W1 = \pi\epsilon^2 E''$$

is the energy dissipated.

Notations used

ω	Angular frequency
k	Dynamic spring stiffness or spring constant
η	Viscosity
c	Damping coefficient
cc	Critical damping coefficient
ζ	Damping ratio
Tr	Transmissibility
g	Acceleration, due to gravity (unit of shock as multiples of g)
$A0$	Peak acceleration for shock
τ_o	Shock pulse duration

References

[1]   V. G. Geethamma, R. Asaletha, N. Kalarikkal, S. Thomas, 'Vibration and sound damping in polymers', *Resonance*, vol. 19, no. 9, pp. 821–833, Sep. 2014, doi: 10.1007/s12045-014-0091-1.

[2]   Ralph E. Blake, 'Chapter 2: Basic Vibration Theory'. Accessed: Nov. 06, 2024 [Online]. Available: *https://www.globalspec.com/reference/64447/203279/chapter-2-basic-vibration-theory*

[3]   P. K. Freakly, A. R. Payne, 'Calculation of natural frequency and damping ratio', in *Theory and Practice of Engineering with Rubber*, Applied Science Publishers Ltd, London, 1978.

[4]   J. E. Ruzicka, 'Passive and Active Shock Isolation'.

[5]   J. E. Ruzicka, 'Active Vibration and Shock Isolation', presented at the National Aeronautic and Space Engineering and Manufacturing Meeting, Feb. 1968, p. 680747. doi: 10.4271/680747.

[6]   T. Irvine, 'An Introduction to the Shock Response Spectrum', vol. Revision S, 2012.

[7]   W. T. Thomson, W. T. Thomson, *Theory of vibration with applications*, 2. ed. London: Allen & Unwin, 1983.

[8]   K. H. Henkel, 'Application of PU for Steering Wheel Production', presented at the SAE International Congress and Exposition, Feb. 1981, p. 810138. doi: 10.4271/810138.

[9]   H. G. Ostfield, 'Polyurethanes in Automotive: Technical Aspects For Their Use in Passenger Vehicles', *J. Cell. Plast.*, vol. 19, no. 3, pp. 141–151, May 1983, doi: 10.1177/0021955X8301900302.

[10] A. Das, P. Mahanwar, 'A brief discussion on advances in polyurethane applications', *Adv. Ind. Eng. Polym. Res.*, vol. 3, no. 3, pp. 93–101, Jul. 2020, doi: 10.1016/j.aiepr.2020.07.002.

[11] N. V. Gama, A. Ferreira, A. Barros-Timmons, 'Polyurethane Foams: Past, Present and Future', *Materials*, vol. 11, no. 10, p. 1841, Sep. 2018, doi: 10.3390/ma11101841.

[12] R. M. E. Diamant, *Thermal and acoustic insulation*. London Boston: Butterworths, 1986.

[13] N. N. Najib, Z. M. Ariff, A. A. Bakar, C. S. Sipaut, 'Correlation between the acoustic and dynamic mechanical properties of natural rubber foam: Effect of foaming temperature', *Mater. Des.*, vol. 32, no. 2, pp. 505–511, Feb. 2011, doi: 10.1016/j.matdes.2010.08.030.

[14] L. Lapcik, M. Vašina, B. Lapčíková, E. Otyepková, K. E. Waters, 'Investigation of advanced mica powder nanocomposite filler materials: Surface energy analysis, powder rheology and sound absorption performance', *Compos. Part B Eng.*, vol. 77, pp. 304–310, Aug. 2015, doi: 10.1016/j.compositesb.2015.03.056.

[15] S. H. Baek, J. H. Kim, 'Polyurethane composite foams including silicone-acrylic particles for enhanced sound absorption via increased damping and frictions of sound waves', *Compos. Sci. Technol.*, vol. 198, p. 108325, Sep. 2020, doi: 10.1016/j.compscitech.2020.108325.

[16] H. Choe, J. H. Lee, J. H. Kim, 'Polyurethane composite foams including CaCO3 fillers for enhanced sound absorption and compression properties', *Compos. Sci. Technol.*, vol. 194, p. 108153, Jul. 2020, doi: 10.1016/j.compscitech.2020.108153.

[17] G. Sung, J. W. Kim, J. H. Kim, 'Fabrication of polyurethane composite foams with magnesium hydroxide filler for improved sound absorption', *J. Ind. Eng. Chem.*, vol. 44, pp. 99–104, Dec. 2016, doi: 10.1016/j.jiec.2016.08.014.

[18] H. Choe, G. Sung, J. H. Kim, 'Chemical treatment of wood fibers to enhance the sound absorption coefficient of flexible polyurethane composite foams', *Compos. Sci. Technol.*, vol. 156, pp. 19–27, Mar. 2018, doi: 10.1016/j.compscitech.2017.12.024.

[19] J.-H. Oh, J. Kim, H. Lee, Y. Kang, I.-K. Oh, 'Directionally Antagonistic Graphene Oxide-Polyurethane Hybrid Aerogel as a Sound Absorber', *ACS Appl. Mater. Interfaces*, vol. 10, no. 26, pp. 22650–22660, Jul. 2018, doi: 10.1021/acsami.8b06361.

[20] S. Basirjafari, R. Malekfar, S. Esmaielzadeh Khadem, 'Low loading of carbon nanotubes to enhance acoustical properties of poly(ether)urethane foams', *J. Appl. Phys.*, vol. 112, no. 10, p. 104312, Nov. 2012, doi: 10.1063/1.4765726.

5

Bridge Bearings and General Elastomer Bearings

Rosamma Alex, Baby Kuriakose, Pradeepkumar P. Joy

Rubber-to-metal bonded blocks are used in many products: bridge-bearing pads and general vibration isolators, such as engine mounts, vibration isolators and suspensions, seismic vibration isolators; torsion blocks such as spheriblocks; flex seals in rockets; and general goods such as engine parts, seals, pedals and conveyor belts. The advantage of laminated bearings is that horizontal stiffness increases while shear stiffness remains almost unaffected, maintaining good flexibility.

In rubber-to-metal bonded blocks, compression is more difficult, because the rubber bonded to the metal cannot bulge freely, resulting in reduced deformation under load. This resistance to deformation is quantified by a term known as the shape factor. The shape factor increases with the number of laminations. Low shape factor designs are used in bridge-bearing pads and seismic vibration isolators.

5.1 Bridge-Bearing Pads

A bearing is a metallic or non-metallic device used extensively in large infrastructure, such as buildings and bridges, to protect them from different movements and vibrations affecting them. For this reason, bearings play a crucial role in the construction sector.

A bridge deck provides a smooth surface for different types of vehicles and pedestrians to travel on. It is generally made of concrete, steel or wood. It helps to distribute the load from the traffic smoothly to the supporting structures. The main horizontal beam that supports the deck is the girder. The primary vertical support for the bridge deck and the girder comes from the piers or beams that extend to the earth.

Generally, the deck and girders are considered part of the bridge's superstructure while the supporting elements, such as abutments, piers or beams, form the substructure. The foundation is considered to be a substructure of the bridge. Bearings are placed between the substructure and the superstructure.

The bridge deck is subject to various movements caused by heavy traffic, thermal expansion and contraction, and environmental factors such as heavy wind and rainfall. In other words, the superstructure can experience horizontal, vertical or rotational movements. Bridges also need protection from vibrations that can arise from structural movements or ground disturbances.

Bridges are designed to accommodate all these movements and loads, with the girders that support the deck resting on piers. Given the heavy load experienced by the deck, there is every chance that piers will get damaged. The load experienced by the deck (superstructure) can be controlled and uniformly transmitted to the piers (substructure) if there is a bearing between the superstructure and the substructure.

Rubber-bearing pads are simple devices that protect the piers by transferring uniform loads from the superstructure to the substructure. In addition to load distribution, they absorb the damaging effects of shock and vibration. They may also reduce noise, absorb compressive forces from the deck and reduce wear and tear between the superstructure and substructure. All these can increase the bridge's life span. The advantages of elastomer bearings are that they combine high compressive stiffness with low shear stiffness. Thus, the function of a bearing is to permit vertical lateral and rotational movements of the superstructure as well as movement of the foundation – mainly to accommodate seismic vibrations.

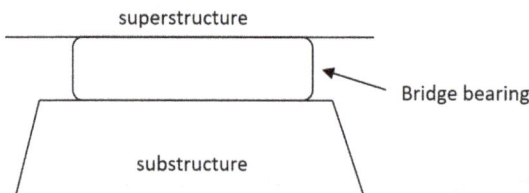

Figure 5.1 Position of a bridge bearing in a bridge

Bridge bearings come in different types, including roller bearings, sliding plate bearings and elastomeric bearings. The last of these offer advantages by virtue of their viscoelastic nature, which allows them to accommodate different types of translational and rotational movements more effectively and over longer periods of service. The benefits include durability, easy installation, almost no maintenance, low cost and versatility for force–deflection characteristics. Elastomeric bearings can accommodate combined deformations, such as shear with rotation or compression with rotation.

A bridge bearing consists of a block of rubber called a plain bearing or a laminated bearing with layers of rubber bonded with steel plates inserted between the bridge deck and support. Natural rubber or chloroprene rubber is widely used to make bridge-bearing pads.

5.1.1.1 Force Deflection Characteristics of Rubber

5.1.1.2 Young's Modulus (E) or Elastic Modulus

Rubber, a soft material, can easily be stretched, compressed or subjected to shear deformation. Under tensile stress, the strain is longitudinal. If a rubber strip of length L and cross-sectional area A is subjected to a deforming force F, its length will increase by a small length δL, as shown in Figure 5.2. A restoring force acts in the direction opposite to the direction of stretch or displacement. Force is linearly related to displacement and the proportionality constant is known as the stiffness K. Stress is proportional to strain.

$$K = F/\delta L = AE/L \tag{5.1}$$

The relationship between tensile stress and tensile strain defines Young's modulus E:

$$E = \frac{\text{Tensile stress } (\sigma)}{\text{Tensile strain } (\varepsilon)} = (F/A)/(\delta L/L) \tag{5.2}$$

Where

E = Young's modulus and its unit is the same as that of stress (lb/inch^2 or N/mm^2)

The value of E is also related to the hardness of rubber. Alternatively, the strip can be subjected to a compressive force.

Compared with steel, rubber has a much lower Young's modulus. For example, E for steel is approximately 30 x 10^6 psi, while for rubber, it may be as low as 100 psi.

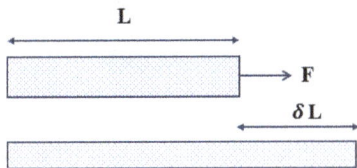

Figure 5.2 Linear deformation of rubber

5.1.1.3 Compressive Stiffness

Consider a rubber block of cross-sectional area A and thickness t under compression (Figure 5.3 a and b). The volume of rubber is assumed to remain constant during deformation, because of its high bulk modulus and a Poisson's ratio of 0.5 [1, 2]. Under compression, the thickness of the block reduces by δt and the rubber bulges at the sides, as shown in Figure 5.3 b.

$$Kc = Fc/\delta t \tag{5.3}$$

$$Kc = EcA/t \tag{5.4}$$

Where

Kc = compressive stiffness

Fc = compressive force

δt = compressive deformation

Ec = compression modulus

A = area of the cross-section

t = original thickness of the block

The compressive stiffness of rubber can be increased by bonding it to thin metal sheets. A block of rubber bonded to the top and bottom surfaces of the metal using suitable adhesive is called a single-layer rubber block. Rubber block with multiple layers of rubber and metal plates (steel plates) is called a multi-layered elastomer bearing. As the number of rubber layers increases, the compressive stiffness of the rubber block increases while the bulged area of the rubber bock decreases, as illustrated in Figure 5.3.

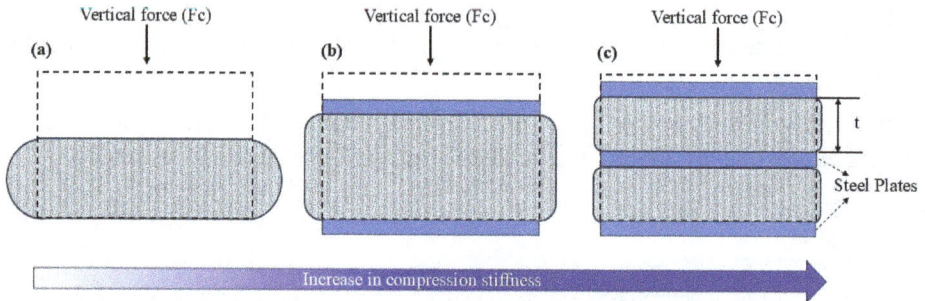

Figure 5.3 Compressive stiffness of rubber bearings (a) without any compressive force, (b) rubber bearing under compression, (c) single-layered rubber bearing with steel plates under compression, (d) double-layered rubber bearing with steel plates under compression

The bulging of the sides decreases as the loaded area increases when rubber surfaces are bonded to metal surfaces. Shear deformations are not affected by inserting metal plates, because shear deformation is related to the deformation of rubber.

The stiffness of rubber under compression when bonded surfaces are prevented from slipping depends on the shape factor S. This is the ratio of one loaded area to the area free to bulge and is a good indicator of compressive stiffness. A high shape factor signifies a rubber of higher stiffness.

For a rectangular block:

$$S = \frac{L \times B}{2t(L + B)} \tag{5.5}$$

For square and cylindrical blocks:

$$S = \frac{L}{4t} \tag{5.6}$$

Where

L = length of the block

B = breadth

t = thickness

Similarly, for circular bearings with a diameter D and thickness t:

$$S = \pi(D/2)^2/2\pi(D/2)t = D/4t \tag{5.7}$$

The shape of the bearing is vital in controlling the shape factor. A circular and square bearing with the same cross-sectional area can have different shape factors. Consider a circular and square elastomeric pad of the same cross-sectional area and same thickness. It will be found that their perimeters will be different and so will the shape factor [3].

The ratio of the shape factor is:

Circular : Square = 1 : 0.88

The area of a circular shape is $\pi D^2/4$, where D = diameter and that of a square shape is A^2, where A = length of one side. Since the area is the same, $A^2 = \pi D^2/4$ and A= $\frac{\sqrt{\pi} D}{2}$

Then the perimeter of the square, $4A = 2 \sqrt{\pi} D$ and the perimeter of the circular shape is πD.

The shape factor is calculated based on the thickness of one rubber layer of a multi-layered block with internal layers of equal thickness. If the thickness of the total rubber layers is T, then the thickness of one layer t =T/n (Figure 5.3 and [4]).

The shape factor measures the variation in the stiffness of bearings with the same cross-sectional area, but different geometries. Compressive modulus can be increased by increasing the shape factor, but the relationship is non-linear and the rate of increase diminishes as the shape factor becomes very large.

In multi-layered elastomer bearings with n layers of elastomer (bonded to steel shims) and each elastomer layer of thickness t, the compressive stiffness Kc is given by:

$$Kc = Fc/\delta t = AL \times Ec/t \times n \tag{5.8}$$

Where

δt = compression displacement,

AL = loaded area

Ec = compressive modulus

The shear stiffness Gs is given by:

$$Gs = Fs/x = AL \times G/T$$

The total thickness of the elastomer layer is:

$$T = n \times t$$

Where

x = Shear deflection

Fs = shear force

G = shear modulus

The compressive modulus E_c is related to Young's modulus E and the shape factor S as follows:

$$E_c = E(1 + 2kS^2) \tag{5.9}$$

Where

k = numerical parameter dependent on the hardness of the rubber

When S > 3, it may be more convenient to use:

$$E_c \approx 5GS^2 \tag{5.10}$$

Where

G = shear modulus

The effective compressive modulus E_c depends on the material property and component geometry since the expandable area depends on the shape of the rubber block.

5.1.1.4 Shear Modulus

When a tangential force (F_t) is applied to a block of cross-sectional area A and thickness T, the sample does not compress or change in length, but instead deforms angularly, forming an angle θ, given by tan θ x/T, due to shear deflection x, as shown in Figure 5.4. The shear modulus G is given by:

$$G = \frac{\text{Shear stress}(\tau)}{\text{Shear strain}(\gamma)} = \frac{F_t/A}{x/T} = \frac{F_t/A}{\tan \theta} \tag{5.11}$$

G is also called the rigidity modulus.

The shear strain is given by tan θ as:

$$\tan \theta = \frac{F_t}{GA}$$

Where

τ = tangential stress

γ = tangential strain

When the value of θ is small,when measured in radians

$$\tan \theta = \theta$$

The shear force, tangential force or horizontal force is F_t:

$$F_t = GAx/T$$

The shear stiffness Gs is then given by:

$$Gs = GA/T \tag{5.12}$$

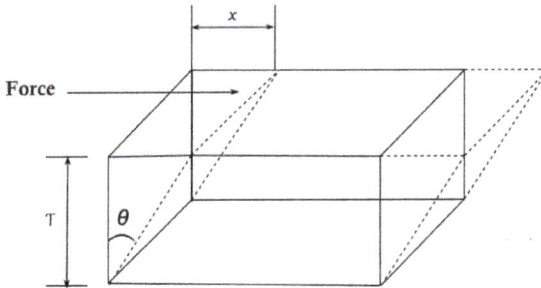

Figure 5.4 Rubber-to-metal bonded rubber block subjected to a shear force

5.1.1.5 Bulk Modulus (K_b)

When stress is applied not in one direction, but in all directions, the sample compresses without changing the shape. The force applied may be hydraulic pressure and the strain may be based on changes in volume. This property is given by bulk modulus K_b, which is defined as hydraulic stress/volume strain.

Bulk modulus:

$$K_b = PV/\delta V \tag{5.13}$$

Where

K_b = bulk modulus

P = pressure applied

$\delta V/V$ = relative change in volume

$$\text{Compressibility} = 1/\text{Bulk modulus} \tag{5.14}$$

5.1.1.6 Poisson's Ratio

Rubber is essentially an incompressible substance. Therefore, during compression or tension, the deformation manifests itself as a change of shape in compression and the sides bulge as shown. When the rubber strip is stretched in one direction, it tends to get thinner in the other two directions, as seen in Figure 5.5.

$$\mu = \varepsilon t/\varepsilon l \tag{5.15}$$

Where

μ = Poisson's ratio

εt = transverse strain

εl = longitudinal or axial strain

Rubber has a Poisson's ratio (lateral strain/longitudinal strain) of approximately 0.5. For materials such as metals, Poisson's ratio is typically lower, at around 0.33.

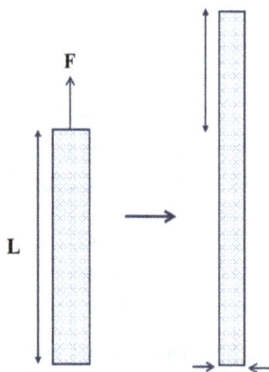

Figure 5.5 Poisson's ratio: μ = lateral strain/longitudinal strain

The following equations relate these elastic constants:

$$E = 2G(1 + \mu) \tag{5.16}$$

E = Young's modulus, G = shear modulus, μ = Poisson's ratio, K_b = bulk modulus.

$$E = 3G$$

$$E = 3K_b(1 - 2\mu) \tag{5.17}$$

5.1.1.7 Stress-Strain Behavior of Rubber

At very low strain, the ratio of the resulting stress to the applied strain is constant and Hooke's law is obeyed in tension and compression (up to about 20% compression). However, as the strain increases, this linearity ceases and Hooke's law no longer applies. Also, the compression and tension stresses are different. In shear deflections, stress is proportional to strain over higher strain ranges.

5.2 Elastomers Used in Bridge-Bearing Pads

Both natural and synthetic rubber can be used. Natural rubber has poor aging characteristics, whereas chloroprene rubber (CR) offers the advantage of good mechanical properties along with enhanced resistance to aging by oxygen, ozone and oils; it is therefore more durable than natural rubber. Chloroprene rubber is preferred to natural rubber for elastomeric pads.

Figure 5.6 Placement of a bridge-bearing pad on a river bridge in Kalpetta, Kerala, India (Courtesy: Public Works Department, Bridges, Kerala state)

Figure 5.7 Three different views depicting the positioning of a bridge-bearing pad on a railway overbridge in Kottayam, Kerala, India

5.2.1 Salient Features of CR

Like synthetic rubbers, chloroprene rubber is also available with different Mooney viscosity values, due to differences in molecular weight and molecular weight-related parameters. High-viscosity grades have higher strength and higher filler loading capacity. The crystallization rate (more trans-structure) can be medium, fast and slow.

Grade with a slow crystallization rate is preferred for elastomeric bearings. Fast crystallization grades are selected for adhesives. Medium crystallization grades are used for hoses and general goods.

With the sulfur-modified grades, there is some polymerization of sulfur onto the main chain in the presence of tetraalkylthiuram sulfide (G grade). The advantage of this grade over mercaptan-modified is that its molecular weight can be reduced during mixing in a mixing mill, thereby making processing easier. Sulfur-modified grades have higher tear strength and better adhesion to steel and fibers than mercaptan-modified ones. However, their heat resistance is inferior. Vulcanization is achieved by using ZnO and MgO. Ethylene thiourea, at a dose of 0.05–1.5 phr, is used in the compound to improve heat resistance. Adding MBTS or TMTD at a dose of 0.1–1 phr is beneficial for low compression strength. High Mooney viscosity grades with higher filler loading are preferred for higher damping applications.

The mercaptan modified (W grades) neoprene composed of poly (2-chloro-1,3-butadiene) units mainly trans configuration and about 10% cis configuration. These grades offer superior dynamic properties over other grades of chloroprene like G grades It has a density of 1.24 at room temperature. At low temperatures and over time, polychloroprene crystallizes and stiffens much more than natural rubber.

Ozone tests recommended by AASHTO (the American Association of State Highway and Transportation Officials) for the elastomeric material used in bridge bearings pad construction are shown in Table 5.1.

5.3 Design of an Elastomeric Bearing: Bridge Bearing

Bridge bearings are rubber blocks bonded to steel plates of specific thickness. Chloroprene rubber-based bridge bearings are designed to meet the requirements of the elastomer and force–deflection characteristics associated with the forces arising in the bridge. The rubber bearing is placed below the superstructure, as shown in the figures below. As elastomeric bearings are used in bridges, the elastomers' aging resistance to the action of oils, weather, atmospheric ozone and extreme temperature is critical. The reaction of rubber with oxygen or ozone can degrade the rubber's elasticity over time. The most commonly used elastomers are natural rubber and polychloroprene rubber, due to their unique force–deflection nature and elasticity. Bearing pads are designed as per international standards such as AASHTO (American Association of State Highway Officials), BS 5400, EN 1337 or DIN 4141 or IRC: 83 (Part II) – 1987: Standard Specifications and Code of Practice for Road Bridges Section: IX Bearings Part II: Elastomer Bearings. According to IRC:83 (Part II), the recommended base polymer is chloroprene rubber (CR) and the detailed specifications of elastomer are provided in Table 5.1.

Table 5.1 Properties of Elastomer for Elastomer Bearings

Parameter	Value	Test Method
Hardness, IRHD	60 ± 5	IS 3400 Part I
Minimum tensile strength, MPa	17	
Minimum elongation at break,%	400	IS 3400 Part II
Compression set; 100 °C for 24 h (maximum)	35	IS 3400 Part X
Accelerated aging; 72 h at 100 °C		
Change in hardness (maximum)	+15	IS 3400 Part IV
Change in tensile strength (maximum)	−15	
Change in elongation (maximum)	−30	

The hardness of rubber varies from 50 to 70 Shore A and is adjusted by adding carbon black. Shear stiffness depends on hardness. As hardness increases, shear stiffness increases. Carbon-black-filled neoprene rubber vulcanizates with IRHD values varying from 50 to 65 exhibit shear modulus G values between 0.64 and 1.37 N/mm^2 [5].

Chloroprene rubber (CR), W grade, offers the best compression set and heat-aging resistance. It requires organic accelerators (ethylene thiourea) for vulcanization. CBS, MBTS and TMTD are effective retarders for delaying vulcanization. CR has inherent ozone and oxidation resistance, but anti-oxidants and antiozonants are required for long-term durability. The details of the AASHTO ozone resistance test for elastomeric bearings are given in Table 5.2. Mixed diaryl para-phenylene diamines and octylated diphenylamine can serve as anti-ozonants and anti-oxidants.

Table 5.2 AASHTO Ozone Test for Elastomeric Bearings

Elastomer Type	Test Condition	Observation
Neoprene rubber	100 pphm ozone in air by volume for 100 hours	No cracks
Natural rubber	25 pphm ozone in air by volume for 48 hours	No cracks

An ozone resistance test was conducted as per ASTM D1149 using test specimens measuring 25 x 150 mm (1 x 6") with a thickness between 1.9 and 2.5 mm (0.075 to 0.10"). The tensile strain was 20% at 38 ± 1 °C (100 ± 2 °F) as determined by ASTM D518, Procedure A.

The fillers are carbon black and silica. Very fine grades of carbon black may be difficult to disperse in a rubber matrix. For good low-temperature flexibility, ester plasticizers are preferred and di-2-ethylhexyl sebacate (also called dioctyl sebacate, DOS) is an effective plasticizer. The rubber compound should have good elastic properties and weather resistance that last several years. A typical formulation based on CR for bridge-bearing compound is given in Table 5.3.

Table 5.3 Typical Chloroprene Rubber-Based Formulation for a Bridge-Bearing Pad

Ingredient	Phr
Neoprene WRT	100
High-activity magnesia (MgO)	4
Octylated diphenylamine	2
Mixed diaryl para-phenylene diamine	2
Microcrystalline wax	0.5
Stearic acid	0.5
ISAF carbon black (N220)	20
HAF carbon black (N330)	10
Di-2-ethylhexyl sebacate	10
Zinc oxide	5
NA22 (ethylene thiourea)	0.7
TMTD	1
Sulfur	0.5

A bearing may consist of an unreinforced plain rubber pad or a reinforced laminated steel plate. Bearings are circular or rectangular and fabricated in suitable lengths, breadths, thickness and area dimensions.

Shearing or a combination of shearing and vertical deformation are common in bridge bearings. Parameters such as permissible shear strain, compressive strain and thickness-to-length ratio are specified in the international standards. Rubber bearings are generally required to have low shear stiffness and high compressive stiffness.

A rectangular elastomeric pad should have high vertical stiffness, low shear modulus, suitable stability based on the a/T ratio (where a is the breadth and T = thickness of the rubber block) and a shear deflection related to the thickness of the rubber. Due to the incorporated steel plates, the laminated bearing supports high vertical loads with minimum deformation, while the flexibility of rubber is maintained under lateral load. Vertical stiffness increases with the shape factor. The thickness of the pad is governed by its shear movement. In designing bridge bearing pads for safety considerations, shear deflection (x) should be less than 0.7 T or the thickness T has to be such that T > 1.43x.

5.4 Rubber-to-Metal Bonding

Special adhesives are used for bonding rubber to low-carbon steel surfaces, such as those used in vibration mounts, during vulcanization. Before the adhesive is applied, the metal surfaces must be thoroughly cleaned to remove oils and greases, corroded materials or other chemical impurities. Oil and grease can be removed by solvent degreasing or alkaline cleaning. Blasting with steel grit or a chemical method of phosphatizing in which a layer of phosphate is formed on the metal surface is also done. Machining, grinding or wire brushing also can be used.

Proprietary bonding agents such as Chemlok Desmodur R serve as rubber-to-metal bonding adhesives. Their exact chemical composition is undisclosed. Early adhesives used cyclized rubber hydrochloride and chlorinated rubber blended with resins and rubber in suitable solvents. Isocyanates were also employed for this. The bonding agents are applied as a one- or two-coat system consisting of a primer layer and an adhesive topcoat to cleaned metal surfaces by brushing or spraying.

Two-coat adhesive systems consist of a primer layer and an adhesive topcoat. A typical primer comprises film-forming polymers, cross-linkable resins, fillers and a solvent or water-borne carrier. The adhesive topcoat is applied after the primer layer is applied and dried. A typical adhesive topcoat comprises curatives, film-forming polymers, fillers and a carrier system.

The primer-to-metal interface involves adsorption and/or chelation of the primer components at the surface of the metal. At the primer-to-adhesive interface, adsorption and diffusion of respective components within the layers occur. The topcoat reacts with active sites in the rubber and primer layers.

Rubber-to-metal parts are manufactured by compression molding. During the molding process, the parameters must be carefully defined so that elastomer vulcanization and adhesive curing occur simultaneously. In addition, the rubber should not have an air gap and there should be intimate contact between the elastomer and adhesive.

Numerical question for the design of a bridge bearing

Design a non-reinforced neoprene rubber based elastomer bearing to suit the following data, based on a plan dimension of 250 x 400 as per a the standard, IRC 83: Part II 1987. Calculate the thickness, shear deflection and compressive stress. Also, test if these parameters are safe for use of the bearing.

- Total vertical force (Ft) = 250 kN

- Horizontal force (Fs) = 50 kN

- Modulus of rigidity (G) = 1.0 N/mm^2

Answer:

Stability for the bearing requires that a/T = 5, where a = width of the bearing and T = total thickness of the bearing

For safety considerations, thickness should be > 1.43 X , lateral deflection or shear deflection (X) should be less than 0.7 T.

The average compressive stress or axial stress σ_a should be $\le$ 2GS, where G = shear modulus and S = shape factor.

Average compressive stress σ_a = Ft/Ae, where Ft = total vertical force, Ae = effective plan area, excluding shear deformation.

Ae = (a – x) b

The dimensions of the bearing is given as breadth (a) = 250 mm and length (b) = 400 mm

For stability of the bearing the thickness (T) can be taken as a/T = 5 , T = 250/5 = 50 mm

Plan area A = breadth (a) x length(b) = 250 × 400 = 100,000 mm^2

$$G = \frac{Fs/A}{X/T} = \frac{X}{T} = \frac{Fs}{GA} = \frac{50 \times 1000}{1 \times 100000} = 0.5$$

$$X = T \times 0.5 = 50 \times 0.5 = 25 \text{ mm}$$

Where

G = Modulus of rigidity

Fs = horizontal force

A = area of cross-section

X = shear deflection

T should be > 1.43X

= 1.43 x 25 = 35.75

which is less than 50 (T) and hence the design is safe.

Shear deflection X should be <0.7 T
0.7 T = 0.7 × 50 = 35
Hence X <0.7 T and the design is safe.

Shape factor = a × b/2T(a + b) = (250 × 400)/2 × 50 × (250 + 400) = 1.538

It is required that σ_a < 2 GS

$$\sigma_a = Ft/Ae$$

Ae = 90,000 mm^2

2GS = 2x 1 × 1.538 = 3.076

$\sigma_a = Ft/(a - X) \times b = 250000/(250 - 25) \times 400 = 2.77$

$\sigma_a < 2\,GS$

$2.77 < 3.076$

So, the design is safe.

5.5 Base Isolation

During an earthquake, the earth's surface shakes in all directions at frequencies ranging from 0.03 to 30 Hz [5]. The strain energy released propagates through the Earth's layers as seismic waves. Consequently, buildings and structures on the ground experience vibrations at their base.

Mathematical expression of base isolation

Consider that there is a displacement of the ground and displacement of the base of a building during an earthquake.

Base isolation is a vibration-isolation technique used for isolating buildings and structures from seismic waves. In this technique, the buildings are isolated from ground movements caused by earthquakes. When buildings or structures are supported on flexible materials called base isolators, they move relatively less than those supported directly on the ground. The earthquake force on a building is a function of a parameter called seismic weight, which is the sum of the seismic weights associated with all floors and the base. Dividing the seismic weight (W) by the acceleration due to gravity yields the seismic mass. Using Newton's equation for force:

Earthquake force

$$F = ma = \frac{W}{g} \times a \qquad\qquad (5.18)$$

The equation can be rewritten as:

$$F = W \times (a/g) \qquad\qquad (5.19)$$

Where

a/g = seismic coefficient

The seismic coefficient, a dimensionless quantity, represents the earthquake-induced acceleration as a ratio of the acceleration due to gravity. It is crucial in seismic design codes for determining seismic loads on structures. For example, if the ground acceleration is 0.5 g, the seismic coefficient is 0.5 and the corresponding earthquake force will be 0.5W, 50% of the seismic weight [4].

Lateral forces predominate in earthquakes. A building undergoing lateral movement must return to its original position afterward and this is made possible by using a material that can store energy and release it after the deforming force is removed. Some energy needs to be dissipated, as all the energy from earthquake forces cannot be stored by the elastic component of rubber.

The principle of base isolation by means of a rubber mount seeks (a) to increase the time period of vibration of the superstructure so that the energy is absorbed for a longer time by the superstructure, (b) to absorb the kinetic energy and dissipate by hysteresis, (c) to reduce the stiffness of the structure.

When considering seismic isolation, it is important to account for the movement of the ground, the base and the building itself, all of which are influenced by the kinetic energy of seismic waves. However, to understand the transmissibility of motion of the superstructure affected by the earthquake, we can model the system as a single-degree-of-freedom base-isolated system, where the building mass is supported by a rubber mount over a vibrating base.

In this system, the base has an input vibration, a rubber mount and a mass (m). The input vibration is from the base, with a displacement y, and this is transmitted to the mass through the vibration isolator, which also vibrates with a displacement x.

Assuming x > y, the relative displacement of the mass is $(x - y)$. The corresponding velocity change is $(\dot{x} - \dot{y})$. Displacement can be evaluated in two ways: as an absolute displacement of the mass (m) or a relative displacement between mass and base.

The forces acting on the mass due to base excitation in the presence of a rubber mount, and the position of the seismic isolator in a building are given in Figure 5.8.

Motion transmissibility is governed by the equation as given below

$$\frac{\text{Amplitude of the structure response}}{\text{Amplitude of the base excitation}} = \frac{X}{Y} \tag{5.20}$$

Where

X = maximum response displacement of structure

Y = input displacement at the input frequency

As given earlier the force due to base excitation is opposed by the spring force, damping force and inertia force of mass in presence of a rubber mount.

In equilibrium condition the sum of different forces acting on mass m, is zero.

$$m\ddot{x} + k(x - y) + c(\dot{x} - \dot{y}) = 0 \tag{5.21}$$

$$m\ddot{x} = -k(x - y) - c(\dot{x} - \dot{y}) \tag{5.22}$$

$$m\ddot{x} + c\dot{x} + kx = c\dot{y} + ky \tag{5.23}$$

The movement of base can be treated as a harmonic vibration and is represented as

$$y = Y \cos \omega t \tag{5.24}$$

$$\dot{y} = -Y\omega \sin \omega t \tag{5.25}$$

Due to the vibration of the base the mass also undergoes harmonic vibration with a difference in phase angle, φ, and is represented as

$$x = X \cos(\omega t - \varphi) \tag{5.26}$$

$$\dot{x} = -X \omega \sin(\omega t - \varphi) \tag{5.27}$$

$$\ddot{x} = -X\omega^2 \cos(\omega t - \varphi) \tag{5.28}$$

On applying these equations in Equation 5.23 and simplifying (see Appendix), we get

$$\frac{X}{Y} = \frac{\sqrt{k^2 + c^2\omega^2}}{\sqrt{(k - m\omega^2)^2 + (c\omega)^2}} \tag{5.29}$$

$$\frac{X}{Y} = \frac{\sqrt{1 + \left(\frac{c\omega}{k}\right)^2}}{\sqrt{\left(1 - \frac{m}{k}\omega^2\right)^2 + (c\omega)^2}} \tag{5.30}$$

Transmissibility in terms of damping ratio (ζ) and frequency ratio ($\frac{\omega}{\omega_n}$),

$$\frac{X}{Y} = \frac{\sqrt{1 + \left(2\zeta\frac{\omega}{\omega_n}\right)^2}}{\sqrt{\left(1 - \left(\frac{\omega}{\omega_n}\right)^2\right)^2 + \left(2\zeta\frac{\omega}{\omega_n}\right)^2}} \tag{5.31}$$

This equation is the same as the equation for force transmissibility as explained in Chapter 4.

The fact that rubber has a very low shear modulus is judiciously exploited in the design of seismic bearings. This is because the shear component of the mechanical forces generated during earthquakes can be efficiently absorbed and dissipated as elastic energy or heat, owing to the inherent hysteresis of rubber, thus protecting heavy structures from damage.

Seismic-resistant bearings are implemented in heavy structures such as bridges and buildings. For instance, a bridge may have a structural component called a shear key, which is a rubber-to-metal bonded component used to prevent lateral movements induced by various climatic conditions, especially seismic waves. The rubber shears easily, even when the compressive stiffness is enhanced by bonding rubber to steel plates. A shear key specifically designed to have a low shear modulus for easy lateral movement is presented in Figure 5.9.

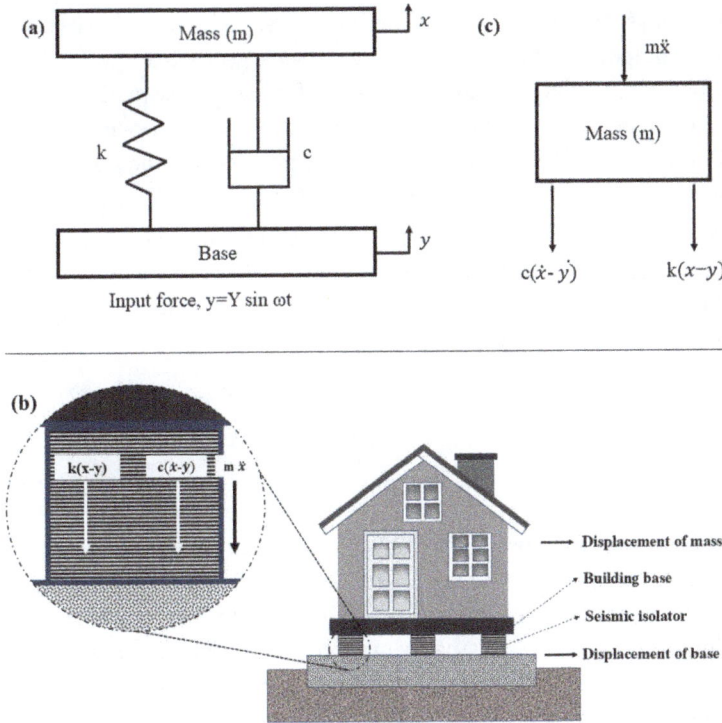

Figure 5.8 (a) Representation of vibration of mass due to base motion in presence of a vibration isolator (b) schematic representation of the seismic isolator in a building (c) free body diagram for the forces acting on mass

Figure 5.9 Shear key used in bridges; top and side views (Courtesy: Hevea Rubber Technologies (P) Ltd, Poovanthuruthu, Kottayam, Kerala)

One of the most effective rubber-to-metal bonded pads developed for base isolators was a rubber steel bonded block with a central lead core. The rubber is viscoelastic by nature and the elastic component can store energy and release it after the deforming force is removed. It may take days, but the deformed rubber will eventually return to its original shape. The viscous component of rubber contributes to energy dissipation. The hysteresis provided viscous component may not be sufficient for energy dissipation. So, to enhance dissipation, an additional material, lead, is introduced into the core of the rubber-to-metal bonded multi-layer block. Lead, a heavy metal, possesses the advantageous quality of being malleable, and can be easily deformed without fracturing.

The rubber provides flexibility through its ability to move, but returns to its original position. If the building has not returned to its original position by the end of the earthquake, the rubber bearings will slowly bring it back. The building can return to its original position, but this may take several months.

5.6 Rail Pads

A rail pad mitigates shock and vibration on a railway track. Also called the sole plate or elastic pad, it is placed between the steel rail and concrete sleepers. Railway rails are designed to withstand vertical forces from the load of the train, lateral forces from the wheelsets and longitudinal forces arising from traction and braking. The rolling wheels generate high-frequency vibrations that can damage the rigid fastening of the rail to the sleeper, followed by wear and tear on the whole track assembly. Vibrations are also transmitted from the rail to the train.

The function of the rail pad is to reduce fatigue cracking of the concrete sleepers and to reduce (by means of vibration isolation or by damping) the vibrations transmitted to the substructure of the railway track. The pads help to distribute the load from the train across multiple sleepers. They protect the sleepers from wear and impact damage and provide electrical insulation for the rails.

Rail pads must be capable of handling high-frequency vibrations through damping and deformation. The pad must possess high stiffness to withstand the load exerted by the train. The use of grooved pads has the advantage of having the required high stiffness to prevent large movements while the wheels pass overhead. Durability is important and the pad should be resistant to environmental conditions. The compression set should be low.

When concrete sleepers were first introduced, thin rubber pads (4.5 mm thickness) were used to protect them. Their service life was relatively low and pads of 6 mm thickness were later developed, with design improvements to secure them in place and prevent slippage.

For the wider concrete sleepers used on high-density routes, 10 mm thick pads are used. The top surface of the rubber pad, which comes directly in contact with the rail, experiences very high-frequency vibrations. In contrast, the bottom surface, in contact with the sleeper, experiences low-frequency vibrations. Rail pads are designed to accommodate the vibrations in these frequencies.

Figure 5.10 Photographs of a rail pad fitted on a track and different types of commercially available rail pads

Natural rubber has high elasticity, while synthetic rubbers such as CR and EPDM have better aging resistance. These rubbers can be used to manufacture rail pads. The rubber compound must have high hardness, good tensile strength, high elasticity and good aging resistance.

5.7 Flex Seal Used in Space Applications

Most satellite launch vehicles and rockets use a flex seal to allow the nozzle to rotate in any direction and act as a pressure-tight seal to allow safe entry into the required orbit. Flex seal contains alternative concentric layers of elastomer pads and shims. The flex seal shown in Figure 5.11 has seven elastomer pads and six steel shims. The essential parts of a flex seal are elastomeric pads, steel shims, fore-end ring and aft-end ring [6]. The type of flex seals determines the thickness of the steel shims, which may vary from 7–10 mm and the elastomer pad may be 3–6 mm thick.

Since the flex seal is a flexible joint, it is made of a natural rubber compound with low hardness. Natural rubber is preferred on account of its low shear modulus and high shear strength. Natural rubber also produces the low compression set, high elongation, strength and high elasticity required for the product and to provide directional change in the propulsion system. The compounds may not contain filler, as the maximum hardness is 40 Shore A and high elasticity needs to be maintained. Natural rubber-based elastomeric pads are the most suitable element for this application, because of the ease with which they can be formulated to give low shear modulus and high shear strength.

The rubber compound must have good extrudability and lend itself to transfer and compression molding. It should also bond well to metal surfaces and carbon fiber-re-inforced epoxy sheets.

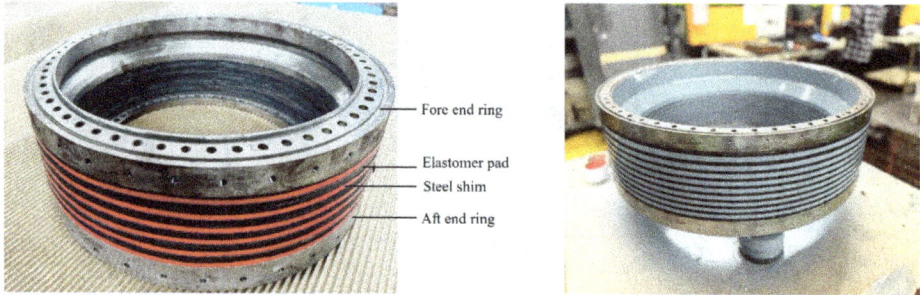

Figure 5.11 Different types of flex seal (Courtesy: M/s Vajra Rubber Products (P) Ltd, Irinjalakuda, Kerala)

5.8 Dock Fender

A dock fender is a shock absorber installed on a wharf so that vessels berth safely without damaging either themselves or the structure. Its function is to absorb the ship's kinetic energy. Old tires were commonly used for this purpose, but rubber dock fenders have since been developed.

The ship's kinetic energy is not determined solely by its mass and velocity. Other contributory factors are the vessel's shape, the portion colliding with the fender and the flow of water moving with the vessel, and so additional parameters are involved in calculating the ship's kinetic energy. A term related to the mass of moving water also needs to be added to the mass of the vessel and is called the added water coefficient. Some of the kinetic energy absorbed by the fender is stored and returned to the vessel, potentially producing a rotational motion, thereby dissipating some of the ship's kinetic energy. A portion of a ship's kinetic energy can be absorbed by the flow of water or *via* the cushioning effect of water trapped between the ship's hull and the wharf; this is known as the configuration coefficient. A ship's hull absorbs some energy after impact and this is referred to as the softness coefficient. All these parameters are considered when calculating the kinetic energy of the ship [7].

Since the kinetic energy is high, the rubber must have a high modulus, good flexibility or elasticity and the ability to absorb kinetic energy by hysteresis. Rubber fenders are generally made from natural rubber or polychloroprene rubber. Dock fenders come in various shapes, with cylindrical fenders being the most commonly used. They have high hysteresis for shock absorption and accommodate the ship's horizontal and vertical movement.

Figure 5.12 Cylindrical dock fenders (Courtesy: M/s Vajra Rubber Products (P) Ltd, Irinjalakuda, Kerala)

5.9 Spheriblock

This is a rubber-to-metal bonded torsion block used in railway engines. There are different types of spheriblocks, which are a component of the axle guide attached to the axle bearing and axil of the engine. The engine's weight is borne by an axle, which is a rod that connects a pair of wheels and transmits power from the engine to the wheels. The spheriblock is fitted to the axle guide to act as a mount for vibration isolation.

5.10 Rubberized Road

Rubberized bitumen in road surfaces provides smoother, safer and quieter road surfaces. Bitumen is a by- product of crude oil distillation. In appearance, it is a highly viscous liquid or solid which is non-volatile and softens on heating. Bitumen is a viscoelastic thermoplastic adhesive applied to road surfaces and it possesses waterproof properties and resistance to aging. Its viscoelasticity lets it absorb strains from moving traffic and temperature fluctuations. Bitumen holds the aggregates firmly together and prevents water entry into the pavement. However, bitumen pavements are susceptible to permanent damage caused by fatigue failure, high-temperature rutting and low-temperature cracking under temperature variations.

Due to the following factors, the dynamic properties and durability are considerably enhanced by modifying bitumen with a small quantity of rubber [8].

- The addition of rubber increases the viscosity and elasticity of bitumen, providing more elastic recovery for rubberized bitumen than ordinary bitumen, reducing high-temperature permanent deformation and mitigating low-temperature thermal cracking, as the bitumen stiffness is reduced. It also reduces fatigue cracking induced by heavy traffic while increasing the load-spreading ability and resistance to distortion.

- Rubberized bitumen in pavements reduces noise, because it allows the formation of larger pores within aggregates, compared with conventional bitumen. Aggregates are the primary materials in road surfacing and are obtained from crushed rocks of various sizes, shapes and textures. Bitumen binds them together. The higher viscosity of rubberized bitumen allows the aggregates to be packed with bigger pores. These bigger pores trap and disperse sound waves, resulting in quieter roads.

- Rubberized bitumen enhances adhesion among aggregates. It improves skid resistance by boosting aggregate retention and eliminating surface bleeding. These combined properties increase the service life of rubberized roads when compared with that of bituminous roads.

Rubberized bitumen is prepared from various forms of rubber: preserved NR latex, unvulcanized dry rubber, vulcanized rubber, shredded reclaimed rubber (used tires) and other vulcanized rubber products.

For latex-based systems, stabilized latex is slowly mixed into bitumen at 140 °C and heated for 30 minutes to reach a final rubber concentration of 2% in the bitumen. When dry rubber is used, it is typically dissolved in a suitable solvent before being added to the bitumen. The interaction of rubber with bitumen involves absorption of bitumen by rubber to an extent depending on the mixing temperature, degree of vulcanization of the rubber and the content of fillers, such as carbon black, in the vulcanized rubber.

5.11 Rubber as Insulating Material

One of the key applications of rubber as an insulating material is that of cables. A cable material must have good insulating properties capability, low dielectric loss, high electrical strength and efficient heat dissipation. Natural rubber was the first polymer used in the electrical industry before the advent of synthetic rubbers and polymers in the early 1930s.

Rubber gloves made from natural rubber latex were first used in medical applications and later for electrical applications to protect against electrical shocks and burns. Gloves can be made from natural rubber latex by a simple process. The general properties of electricians' gloves made from natural rubber are given in IS 4770: 1991. There are four types of gloves, based on voltage levels.

- *Type 1* – For use at a voltage not exceeding 650 AC rms

- *Type 2* – For use at a voltage not exceeding 1100 AC rms

- *Type 3* – For use at a voltage not exceeding 7500 AC rms

- *Type 4* – For use at a voltage not exceeding 17,000 AC rms

Compliance with the standard requires the gloves to meet criteria for thickness, tensile strength, ultimate elongation, puncture resistance, moisture absorption, proof voltage, leakage current and breakdown voltage [9].

The glove material should exhibit high strength and elasticity, with minimal non-rubber ingredients that may impair electrical resistivity or affect related parameters, such as DC breakdown voltage and current leakage. The material should also have ozone resistance.

Two key issues related to the presence of proteins in natural rubber latex film are water absorption, which can result in current leakage, and the potential for allergic reactions arising from leachable proteins. Proteins can be eliminated through the use of double-centrifuged latex, deproteinized latex and appropriate leaching during both the wet-gel and post-vulcanization stages.

Good modulus for the glove can be obtained through the use of ZMBT along with ZDC as accelerators (a typical formulation is given Table 5.4). A good level of vulcanization is important for obtaining sufficient strength-related properties. The thickness and size of the gloves should comply with international standards.

The latex used in electricians' gloves is double-centrifuged or deproteinized, because proteins and other non-rubber ingredients normally present in single-centrifuged latex can adversely affect the insulation quality and are therefore removed. A suitable leaching process is employed to remove the ionic impurities likely to be formed during processing.

Table 5.4 Formulation for Electricians' Gloves

Ingredient	Phr	Wet Weight
60% Double-centrifuged natural rubber latex	100	167
10% KOH solution	0.1	1
50% Zinc oxide dispersion	0.5	1

Ingredient	Phr	Wet Weight
50% ZDC dispersion	1.0	2
50% ZMBT dispersion	1.0	1
50% Anti-degradant HS dispersion	1.5	3
50% Sulfur dispersion	2.0	4

The gloves are manufactured in a conventional dipping process with appropriately shaped formers. The steps involved are dipping, wet-gel leaching, vulcanization, post-leaching, drying and testing. The formers are heated to about 50 °C and dipped in a suitable coagulant to form a coating of the coagulant. The coagulant is 20% aqueous formic acid or 50% calcium nitrate in an alcohol-water mixture. After coating, the formers are dipped into the compounded latex (Table 5.4). A dwell time of about four minutes before withdrawal ensures that a latex film of sufficient thickness is obtained. A latex gel is produced by dipping in the coagulant. Once the wet gel has partially dried, the process is repeated to build up the required thickness. The wet gel is then leached in water. Once dried, the glove is stripped from the former and vulcanized by heating at 100 °C for about one hour or longer, depending on the glove's thickness, to ensure proper vulcanization. The vulcanized glove is then subjected to a post-leaching process by immersion in water for about 24 hours. Finally, the glove is tested to verify compliance with the specifications and standards.

5.12 Anti-Static Gloves

In certain environments, static electricity can accumulate on surfaces, often as a result of friction between two dissimilar materials – such as hair and a plastic comb or the human body and woolen clothing – as well as through the piezoelectric and pyro-electric effects. This build-up of static charge can lead to electric shocks. The heat or spark of a static electricity discharge also can also ignite materials such as oils, organic solvents and flammable materials. The static electricity discharge causes considerable damage to electronic equipment, because the sudden burst of energy produces heat, hence damaging the delicate electrical circuit and circuit boards. Static discharge poses risks to objects with static charge build-up, especially electronic equipment. Both the equipment and individuals working with the equipment may suffer electric shocks, due to the flow of electric current.

Anti-static materials are designed to prevent the rapid build-up of static charges by allowing slow and controlled dissipation. Anti-static gloves are manufactured in line with international standards EN 16350 and IEC 61340. EN 16350–2014 specifies that the glove should have a vertical or volume resistance less than 1 x 10^8 ohms. These gloves are designed with additives such as conductive carbon black to provide a path for the flow of static charges and also to slowly dissipate the charges to the surrounding environment or other surfaces. Owing to this property, anti-static gloves are also called electrostatic discharge (ESD) gloves. Anti-static gloves are used in conjunction with anti-static wrist straps, appropriate clothing and footwear to facilitate the safe discharge of static electricity.

Materials with low electrical resistance, typically between 0.1 and 1000 megaohms, dissipate static charges instead of allowing them to accumulate. Anti-static mats, commonly used in electronic workspaces, operate on this principle. Rubber mats embedded with conductive fillers such as carbon black effectively channel static electricity to the ground, where it is neutralized. Similarly, anti-static footwear dissipates static charges by conducting them safely to the ground.

Dry box gloves, widely used in medical applications, must be both anti-static and resistant to environmental degrading agents, such as oxygen, ozone and UV radiation. They are typically made from EPDM or butyl rubber, although natural rubber is also an option. ESD-safe gloves contain fillers such as carbon black to facilitate the slow dissipation of static charges and they usually have a resistance range between 10^5 and 10^8 ohms. Conductivity levels below 10^5 ohms can lead to a rapid charge flow, which may be hazardous.

Natural rubber, EPDM, nitrile and butyl rubber are all suitable for making anti-static gloves. Latex dipping is commonly used for rubber in latex form, while solution dipping is preferred for EPDM-based gloves. Conductive fillers such as carbon black are frequently added to rubber on account of their ease of dispersion, cost-effectiveness and availability. However, other fillers such as nanoclays, carbon nanotubes and metal flakes can also be used to improve the conductivity of rubber.

Appendix

Mathematical equations involved in motion transmissibility

Transmissibility (Tr)

$$\frac{\text{Amplitude of the structure response}}{\text{Amplitude of the base excitation}} = \frac{X}{Y}$$

Equations for the harmonic vibration of base with amplitude y and that of mass with amplitude x is given below

$$y = Y \cos \omega t$$

$$\dot{y} = -Y\omega \sin \omega t$$

$$x = X\cos(\omega t - \varphi)$$

$$\dot{x} = -X\,\omega\sin(\omega t - \varphi)$$

$$\ddot{x} = -X\omega^2\cos(\omega t - \varphi)$$

The general differential equation for base isolation using rubber mount consisting of damper and spring elements is given below

$$m\ddot{x} + c\dot{x} + kx = c\dot{y} + ky$$

Substituting for $x, \dot{x}, \ddot{x}, y \text{ and } \dot{y}$ in the differential equation,

$$-mX\omega^2\cos(\omega t - \varphi) - cX\,\omega\sin(\omega t - \varphi) + kX\cos(\omega t - \varphi)$$
$$= -cY\omega\sin\omega t + kY\cos\omega t$$

$$-mX\omega^2(\cos\omega t\cos\varphi + \sin\omega t\sin\varphi) - c\,X\,\omega(\sin\omega t\cos\varphi - \cos\omega t\,\sin\varphi)$$
$$+kX(\cos\omega t\cos\varphi + \sin\omega t\sin\varphi) = -cY\omega\sin\omega t + kY\cos\omega t$$

Comparing the terms of $\cos\omega t$, we get

$$-mX\omega^2\cos\varphi + cX\,\omega\,\sin\varphi + kX\cos\varphi = kY$$

$$[(-m\omega^2 + k)\cos\varphi + c\omega\sin\varphi]X = kY$$

Comparing the terms of $\sin\omega t$, we get

$$-mX\omega^2\sin\varphi - cX\,\omega\cos\varphi + kX\sin\varphi = -cY\omega$$

$$[(-m\omega^2 + k)\sin\varphi - c\omega\,\cos\varphi]X = -cY\omega$$

On adding and squaring the two equations, we get

$$[(-m\omega^2 + k)^2 + c\omega^2]X^2 = Y^2(k^2 + c^2\omega^2)$$

$$\frac{X}{Y} = \frac{\sqrt{k^2 + c^2\omega^2}}{\sqrt{(k - m\omega^2)^2 + (c\omega)^2}}$$

$$\frac{X}{Y} = \frac{\sqrt{1 + \left(\frac{c\omega}{k}\right)^2}}{\sqrt{\left(1 - \frac{m}{k}\omega^2\right)^2 + (c\omega)^2}}$$

$$\frac{X}{Y} = \frac{\sqrt{1 + \left(2\zeta\frac{\omega}{\omega_n}\right)^2}}{\sqrt{\left(1 - \left(\frac{\omega}{\omega_n}\right)^2\right)^2 + \left(2\zeta\frac{\omega}{\omega_n}\right)^2}}$$

Notations used

E	Young's modulus
k	Linear stiffness
δL	Change in linear deformation
σ	Tensile stress
ε	Tensile strain
Fc	Compressive force
Kc	Compressive stiffness
Ec	Compression modulus
Δt	Change in compression
Fs	Shear force
G	Shear modulus
Gs	Shear stiffness
δx	Shear deflection
S	Shape factor
μ	Poisson's ratio
K_b	Bulk modulus

References

[1]   A. N. Gent, 'On the Relation between Indentation Hardness and Young's Modulus', *Rubber Chem. Technol.*, vol. 31, no. 4, pp. 896–906, Sep. 1958, doi: 10.5254/1.3542351.

[2]   P. B. Lindley, *Engineering design with natural rubber.*, 4th ed. Brickendonbury: Malaysian Rubber Producers Research Association, 1974.

[3]   Niyazi Özgür Bezgin, 'Effects of shape factor on the behavior of elastomeric roadway bridge bearings and benefits of circular bearing cross section', *Chall. J. Struct. Mech.*, 2015, doi: 10.20528/cjsmec.2015.09.032.

[4]   Standard Specifications and Code of Practice for Road Bridges SECTION : IX BEARINGS Elastomeric Bearings Part II 2nd revision IRC: 83–2018 (Part II)

[5]   C. V. R. Murty, 'Learning Earthquake Design and Construction', *Natl. Inf. Cent. Earthq. Eng.*.

[6]   C. Jagadish, Ch. Siva Rama Krishna, P. Abhishek, 'Harmonic Analysis of Flex Seal of Rocket Nozzle', *IOP Conf. Ser. Mater. Sci. Eng.*, vol. 1185, no. 1, p. 012031, Sep. 2021, doi: 10.1088/1757-899X/1185/1/012031.

[7]   W. Galor, 'The Analysis of Effective Energy of Ship's Berthing to the Quay', *J. KONES Powertrain Transp.*, vol. 19, no. 4, pp. 185–191, Jan. 2015, doi: 10.5604/12314005.1138340.

[8]   M. G. Krishnapriya, 'Performance Evaluation of Natural Rubber Modified Bituminous Mixes.', *Int. J. Civ. Struct. Environ. Infrastruct. Eng. Res. Dev. IJCSEIERD*, vol. 5, pp. 121–134, 2015.

[9]   Bureau of Indian Standards, *IS 4770: Rubber Gloves - Electrical Purposes*. 1991. Accessed: Nov. 13, 2024 [Online]. Available: *http://archive.org/details/gov.in.is.4770.1991*

Index

Symbols

4,4′-oxybis-benzenesulfonyl-
hydrazide *188*
α-terminal *4*
ω-terminal *5*

A

abnormal groups *83, 84*
abrasion resistance *101, 102, 104, 105,
108, 110, 136*
accelerated aging *25, 41*
accelerator *14, 15*
acetic acid *10, 27, 30, 40*
acid number *69, 70*
acoustics *182*
activation energy *129, 132, 142*
activator *14*
agglomerated fillers *67*
aging *57, 58, 62, 65, 86*
aging resistance *205, 206, 215*
air-dried sheet *27*
air impermeability *58*
amino acids *5, 7, 11*
ammonia *9, 16, 17, 18, 26, 28*
ammonium carbonate *188*
ammonium oleat *21*

angular

angular displacement *152, 156*
angular frequency *120*
angular velocity *152, 156*
anionic surfactant *68*
anti-coagulants *16, 28*
anti-oxidants *6*
anti-static gloves *62, 220*
Arrhenius equation *99, 129, 132, 141*
audible frequency range *182*

B

Bagley correction *89, 90*
Banbury *64*
bearing *195, 198, 207*
Bingham plastic fluids *54*
Bitumen *217, 218*
bleaching *40*
blowing agent *186, 188*
Brabender *92, 95*
branching *82, 84*
breakdown voltage *219*
bridge bearing *197, 208*
bridge-bearing pads *195, 197*
bulk modulus *197, 201, 202, 203*
buoyancy *19, 20, 42, 44*
butyl rubber *58, 169, 174*

C

calendering   61, 63, 75, 76
capillary rheometer   88
carbon black   13, 50, 57, 61, 66, 68, 69,
101, 103, 104, 105, 106, 107, 110, 115, 119,
141, 173, 206, 207, 218, 221
car shock absorber   156
cellular rubber   185
centrifuge   21, 22, 24
centrifuged latex   8, 11, 15, 22, 24, 25, 34
centrifuging   7, 18, 20, 21, 22, 24
chain branching   96
chain entanglements   99, 113
chloroform number test   14
chloroprene rubber   197, 203, 204, 205
cis-1,4-polyisoprene   6
coagulation   9, 10, 31, 40
– preferential   40
– quick, of latex   12
– spontaneous   5
– types   12
coefficient of friction   137
coefficient of thermal expansion   73
colloidal stability   5, 6, 7, 8, 11, 12, 13, 17
colloidal stabilization system   68
complex modulus   121, 122
compressive stress   208, 209
concrete sleepers   214
continuous mixers   50, 64
creaming agents   18
creaming process   18, 19
creep   100, 101, 110, 111, 112, 113, 123, 126,
127, 128
Crepe rubber   39
creping machine   34
critical damping   156, 167
critical damping coefficient   157
cross-links   15, 41
current leakage   219
cytoplasmic serum   3

D

damping   148, 149, 154, 161, 169
– external factors   150
– forces   153
– types   154
damping capacity   167
damping coefficient   154, 155, 166
damping force   164
damping ratio   149, 157, 166
dashpot   154, 155
De Mattia Flex Tester   117
depletion forces   8
deproteinized   219
deproteinized latex   7, 15
diammonium hydrogen phosphate   20
dibutyl phthalate   61
dielectric loss   218
die swell   57, 66, 72, 74, 85, 90, 95, 96
dilatant fluids   54
dioctyl phthalate   61
dipping process   220
discoloration   30, 32, 33, 40
dispersion   49, 61, 65, 67
dispersion analyzer   70
displacement vector   164
disturbing frequency   168, 170
dock fender   216
double-centrifuged   219
double-centrifuged latex   25
dry box gloves   221
dry rubber content   4, 17, 22, 28, 30, 40, 44
Dunlop method   186
Dunlop tripsometer   118
dynamic properties   172
dynamic stiffness   161, 172

E

earthquake force   210
elasticity   205, 215, 216, 218, 219
elastic modulus   116, 121, 122, 123, 128, 143
electrical applications   219
electrical double layer   8

electrical strength   218
electric potential   8
electrodecantation   25
electrostatic discharge (ESD) gloves   221
electrostatic forces   8
engine mounts   171, 195
enzymatic discoloration   30, 40
EPDM   58, 61, 62, 79
EPDM rubber   171
estate brown crepe   16, 27, 39, 41
ethylene oxide condensate   9
ethylene thiourea   206, 207
evaporation   18, 25
expanded rubber   187
external excitation   150
extrusion   50, 61, 63

F

factice   74
fatigue cracking   214, 218
fatty acid soap   5, 11, 13, 14, 68, 186
fibers   183, 184, 190
field coagulum   16, 27, 34, 37, 39, 41
field latex   3, 9, 17, 19, 28, 30, 39
filler aggregates   65, 66
filler dispersion   2, 57, 61, 65, 67, 96
filler–filler interactions   65, 68, 89
filler masterbatch   68
filler–rubber interactions   59, 68
fillers   49, 57, 61, 64, 66, 67
– proportion   73
flex seals   195, 215
fluid viscosity   49, 51, 53
foam   148, 187, 189, 190
foam stabilizer   186
formic acid   10, 11, 13, 27, 30, 40, 44, 68, 69
free radicals   66
free-vibration system   151
free volume   99, 110, 128, 129, 130, 132
fresh field latex   12
Frey-Wyssling particles   3
furnace blacks   107

G

gamma radiation   7
Garvey die head   91
gelation   9
gelation of latex   187
gel content   83
gel formation   82, 84
glass transition region   129
glass transition temperature (Tg)   99, 102
glassy region   129, 132
Goodrich flexometer   117
graphitic structures   106
grooved pads   214
gum strength   81

H

hammer mill   35, 37
hardness   197, 200, 206, 215
heat build-up   117
Hevea brasiliensis   1, 2, 3
Hobart planetary mixer   186
Hooke's law   203
horizontal force   201, 209
horizontal stiffness   195
hydraulic press   37
hydrodynamic effect   59, 105
hydrofluoric acid   187
hysteresis   100, 115, 129

I

Indian Standard Natural Rubber   37
initial plasticity P0   11, 85
injection molding   72
insulating material   218
intermeshing rotors   50, 64
intermix   64
internal mixer   59, 67
internal mixers   50, 64, 66
ionic strength   8
isocyanates   189
isoelectric point   6, 10, 17

K

kinetic energy   *147, 148, 174, 178, 180, 211, 216*
kinetic-energy-storing device   *149*
kneaders   *50, 64*

L

laminated bearings   *195*
latex allergy   *6*
latex coagulation   *3*
latex concentrate   *4, 15, 16, 17, 21, 24, 25*
latex filler masterbatches   *12*
latex foam   *186*
latex masterbatch   *140*
latex putrefaction   *5, 16*
latex stabilizer   *9, 18*
L/D ratio of extruder barrels   *73*
leaching   *7, 15*
loss factor   *160, 169*
loss modulus   *133*
low-protein latex   *7*
low shear modulus   *212, 215*
L-quebrachitol   *4*
Lupke pendulum   *118*
lutoids   *3, 11*

M

macrostructure of natural rubber   *83*
magnesium ammonium phosphate   *17, 20*
magnesium ions   *12, 20*
Malvern Zetasizer   *14*
marketable forms   *2, 10, 16, 34, 42*
mastication   *65, 66, 67*
– mechanical   *95*
material damping   *158*
Maxwell model   *100, 123, 126, 132*
mechanical stability   *11, 25*
Metrolac   *29, 30*
MgO   *205, 207*

M (continued)

microwave oven drying   *30*
microwave vulcanization   *76, 79*
minimum initial plasticity P0   *64*
mixing equipment   *59*
mixing mill   *64, 67, 92*
modified coagulation process   *68*
modulus   *101, 103, 104, 105, 106, 108, 110, 111, 112, 115, 120, 121, 129, 140*
mold growth   *33*
molding   *50, 61, 63*
molding pressure   *73*
molecular weight   *6, 41, 49, 57, 65, 66, 80, 82, 84, 85, 87, 95*
molecular weight distribution   *6*
Mooney viscometer   *86, 87*
Mooney viscosity   *64, 65, 69*
moving-die rheometer   *87, 88*
Mullins effect   *111, 119*

N

naphthenic oil   *61, 71*
natural angular frequency   *153, 157*
natural vibration   *150, 152*
Newtonian fluids   *52, 53, 54, 89*
noise-damping   *190*
non-Newtonian   *53, 56, 57, 89*
non-Newtonian behavior   *59*
non-rubber ingredients   *1, 2, 4, 12, 15, 25, 27*

O

open mills   *50, 59, 64, 65*
organic peroxides   *62*
oscillating disk rheometer   *87*

P

pale latex crepe   *16, 27*
paraffin oil   *61*
Parthenium argentatum   *1*
particle size distribution   *15, 18, 21, 24*

Payne effect   *119*
permanent set   *111, 113*
phase angle   *158*
phase lag   *100*
phenolic resins   *62*
phospholipids   *4, 5, 8, 10, 24*
pine tar   *61*
pinhead bubbles   *32*
plasticity retention index PRI   *6, 85, 86*
plasticizers   *49, 57, 61, 66*
Poisson's ratio   *197, 202*
polyols   *189*
polyurethane (PU) foams   *189*
pre-breaker   *34, 36*
pre-coagulation   *40*
preservatives   *7, 16, 17, 18*
preserved latex   *4, 8, 10, 15, 25*
pre-vulcanization   *7*
pre-vulcanized latex   *14*
PRI   *70*
pro-oxidants   *6, 17*
proteins   *2, 4, 6, 10, 15, 16*
– water-soluble   *7*
proteolytic enzymes   *7*
pseudoplastic   *53, 54, 55*
p-toluene sulfonyl hydrazide   *188*

R

Rabinowitch correction   *89*
rail pad   *214, 215*
reclaim rubber   *42*
reinforcing fillers   *57, 59, 61, 65, 74, 82,*
99, 104, 106, 108
relaxation modulus   *100*
relaxation time   *125, 126*
resorcinol formaldehyde latex   *60*
restoring force   *151, 197*
retardation time   *127, 128*
rheology   *49, 57, 84*
rheopectic   *55, 56*
ribbed smoked sheets   *10, 16, 27, 28*
rigidity modulus   *201*

rolling resistance   *110, 136, 141*
rubber elongation factor   *10*
rubber–filler interactions   *65, 67, 99, 106,*
107, 108, 120
rubber hydrocarbon   *2, 3, 4*
Rubberized bitumen   *217, 218*
rubber particles   *2, 4, 8, 19, 21, 24, 44*
– negatively charged   *25*
rubber sponge   *187*
rubber steel bonded block   *213*
rubber-to-metal bonded blocks   *195*
rubbery region   *129, 132, 139*
Russian dandelion   *1*
Rust   *33*

S

SBR latex   *60*
seismic coefficient   *210*
seismic vibration isolators   *195*
seismic waves   *210, 211, 212*
seismic weight   *210*
self-vulcanizate rubber blends   *62*
serum   *18, 19, 44*
– aqueous   *44*
shape factor   *195, 199, 200, 207, 209*
shear deformation   *50, 95, 197, 198, 209*
shear forces   *50, 56, 59, 64, 65, 66*
shear key   *212*
shear modulus
– low   *207*
shear rate   *53, 54, 61, 73, 88, 89*
shear stiffness   *195, 196, 200, 201, 206, 207*
shear thinning   *53*
shift factor   *131*
shock   *147, 150, 169, 178, 181*
shock absorber   *157, 181*
– automotive   *154*
shock excitation   *175*
shock output   *179*
shock pulse   *176*
shock transmissibility   *180*
shredder   *35, 37*

silane coupling agent   67, 109
silanol groups   109
silica   13, 101, 103, 104, 105, 108, 109, 119, 140, 141, 173
silicone rubber   58, 171, 174, 187
simple harmonic motion   152, 155
sinusoidal deformation   100, 101, 110, 117, 120, 121, 122, 123, 135
skim   22
skim latex   19, 21, 22, 24
skim rubber   24
skim screw   21, 22, 24
slab cutter   34
sludge   17, 20, 22, 25
sodium alginate   20
sodium bicarbonate   186, 188
sodium metabisulfite   30, 40
sodium silicofluoride   9, 186
sodium sulfite   16, 28, 33, 40
softeners   61
sole plate   214
sound   147
sound-absorbing materials   182
sound absorption   148, 184, 187, 190
sound power   184
specific volume   130, 131
spheriblock   217
sponge   148
sponge rubber   185
spontaneous coagulation   12
spring-mass system   151, 153, 154, 156, 179
static deflection   153, 178, 181
static electricity   220, 221
steric stability   8
stiffness   197
– compressive   196, 198
– variation in   199
– vertical   207
Stokes formula   20
storage hardening   83, 84
strain amplification   59, 105, 119
strain energy   115, 160, 178, 185
strain-induced crystallization   81, 82

stress relaxation   100, 110, 111, 113, 123, 125, 126, 132, 133, 142
– chemical   111
– comparative   115
stress relaxation modulus   125
substructure   196, 214
sulfur   7, 12, 14, 15
superior processing rubber   58
superstructure   196, 205, 211
surfactant   10, 13, 14
synthetic polyisoprene   82
synthetic rubbers   58, 80, 95

T

tandem mixer technology   64
tangential force   200, 201
tangential mixer   64
tangential rotors   50, 64
tapping   2, 12, 16, 28
Taraxacum koksaghyz   1
tear strength   101, 103, 104, 108, 142
technically specified rubber   10, 16, 27
temperature-induced crystallization   81
temperature shift factor   133
tensile strength   101, 103, 104, 142, 206, 215, 219
tetraethylenepentamine   85
tetramethylthiuram disulfide   18
thermal blacks   108
thixotropic   55, 56
time–temperature equivalence   100, 134
time-temperature superposition   141
torsion block   217
traction   136, 137
transmissibility   149, 162, 166, 167, 168, 180
transmitted force   165, 180
triallyl cyanurate   62
turgor pressure   3
two-roll mixers   50, 64

U

underdamped   156, 157

V

Van der Waals forces   *8*
vegetable oils   *61*
velocity shock   *175*
vibration damping   *171*
vibration isolators   *161, 180, 195*
vibrations   *195, 196, 210, 214*
vibratory system   *149, 154, 155, 162, 164*
vinyl pyridine latex   *60*
viscoelasticity   *100, 102, 115, 116, 136*
viscoelastic material   *49, 56, 63*
viscoelastic nature   *196*
viscosity   *25, 39, 44, 49, 52, 53, 54, 65, 66, 85*
– dynamic   *87*
viscous drag   *19, 42, 44*
Voight model   *100, 123, 128*
volatile fatty acids   *10, 12*
vulcanization   *6, 14, 15, 16, 24, 27, 41, 99, 101, 103, 105, 107, 108, 109, 110, 139*
vulcanizing agent   *62*

W

Wallace rapid plastimeter   *85*
Weissenberg effect   *75*
WLF equation   *129*

Y

Yerzley Mechanical Oscillograph   *118*
Young's modulus   *123, 143, 197, 200, 203*

Z

zeta potential   *8, 13, 14, 17*
zinc–amine complexes   *9*
zinc oxide   *9*
ZnO   *173, 187, 188, 205*